# Green Energy and Technology

More information about this series at http://www.springer.com/series/8059

Laura Maturi · Jennifer Adami

# Building Integrated Photovoltaic (BIPV) in Trentino Alto Adige

With Contributions by

Francesca Tilli · Philip Ingenhoven
Marco Lovati · Stefano Avesani · Alessio Passera
David Moser · Roberto Lollini

 Springer

Laura Maturi
Eurac Research
Bolzano
Italy

Jennifer Adami
Eurac Research
Bolzano
Italy

ISSN 1865-3529          ISSN 1865-3537   (electronic)
Green Energy and Technology
ISBN 978-3-319-89269-6          ISBN 978-3-319-74116-1   (eBook)
https://doi.org/10.1007/978-3-319-74116-1

Printed on acid-free paper

This Springer imprint is published by the registered company Springer International Publishing AG part of Springer Nature.
The registered company address is: Gewerbestrasse 11, 6330 Cham, Switzerland

# Preface

This publication is elaborated in the framework of the experience gathered on the BIPV topic by the Institute for Renewable Energy of Eurac within two international projects, i.e. SHC IEA Task 41 "Solar Energy and Architecture" and PVPS IEA Task 15 "Enabling Framework for the Acceleration of BIPV", which was co-financed by the Stiftung Südtiroler Sparkasse.

GSE (Gestore Servizi Energetici) has kindly provided contents for the publication, giving an overview on the BIPV status in Italy through the Feed-in tariff evolution.

The goal of this book is to present inspiring, informative and appealing projects with solar energy for the general practicing architect, designer and developer, by showing exemplary selected projects realized in "Trentino Alto Adige". This region has been very active in recent years in the BIPV field by boosting PV use and building energy efficiency through several measures: incentive schemes, dedicated policies, awareness raising, guidelines development and public engagement in the use of PV in public buildings.

In order to collect the best and most representative case studies present in the "Trentino Alto Adige" region, a "call for case studies" was launched by contacting most of the engineers, architects and professionals of the region.

Out of more than 40 collected cases, the best ones, presented in this book, were selected through an internal workshop involving 19 EURAC collaborators, forming a "selection group". The authors would like to thank EURAC colleagues involved in this selection group (Stefano Avesani, Roberto Lollini, David Moser, Marco Lovati, Matteo Del Buono, Alessio Passera, Adriano Bisello, Dagmar Exner, Francesca Roberti, Laura Scafidi, Cristiana Losso, Matteo Prina, Giulia Paoletti, Ramon Pascual, Monika Mutschlechner, Grazia Barchi, Marco Pierro).

The authors would also like to thank the experts who were interviewed and kindly provided pictures and information for the case studies: Augustin Clement (Ütia da Ju owner), Marina Gemmi (Ertex Solartechnik GmbH), Eng. Matteo Ruzza (Electro Clara Sas), Don Vijo Luigi Alois Pitscheider (Milland priest), Günter Richard Wett, Eng. Klaus Fleischmann (Fleischmann Dr. Klaus & Janser Dr. Gerhard), Eva Pichler (Colterenzio Winery), Arch. Gerd Bergmeister (bergmeisterwolf), Single-family house owner, Eng. Giuseppe Cosentino (Ing.studio Blasbichler Co. Ltd), Florian Maurer (Leitner Electro Srl), Eng. Norbert Klammsteiner (Energytech Srl), Arch. Armin Pedevilla (Pedevilla), Arch. Fabio Rossa (Area 17), Multi-family house owner, Lukas Goller (PVEnergy), Phys. Francesco Nesi (ZEPHIR Srl), Tiziano Maraner (FAR Systems SpA), Eng. Andrea Giacomelli (Autobrennero SpA), Arch. Michael Tribus, Günther Schuster (Elpo GmbH), Eng. Francesco Besana (FAR Systems Srl), Arch. Norbert Dalsass (Arch Panta Rei), Andreas Brunner (Maso Lampele owner), Werner Graber (Elektro Josef Graber), Günther Wallnöfer (Wallnöfer Günther E. Rudolf Snc), Arch. Gianluca Perottoni, Federico Cesaro (Schüco International Italia Srl), Arch. Wolfgang Simmerle, Michael Reifer (Frener & Reifer GmbH), Marlis Schenk (Obrist GmbH), Eng. Norbert Klammsteiner (Energytech Srl), Arch. Albert Colz, Gerhard Strobl (Elektrostudio). Moreover, thanks to Ordine degli Architetti di Bolzano and Ordine degli Ingegneri di Bolzano for supporting the call for case studies.

Bolzano, Italy                                                                    Laura Maturi
                                                                                Jennifer Adami

# Contents

# Chapter 1
# Context and BIPV Concept

**Abstract** The building sector accounts for over 40% of the European total primary energy use and 24% of greenhouse gas emissions (OECD/IEA and AFD, Promoting energy efficiency investments, 2008, [1], IEA, Task 40—towards net zero energy solar buildings, 2009, [2]). A combination of making buildings more energy efficient and using a larger fraction of renewable energy is, therefore, a key issue to reduce the non-renewable energy use and greenhouse gas emissions. It is an important objective of energy policy and strategy in Europe and Italy. In this context, the photovoltaic sector has seen a great development in the past few years, also thanks to incentive schemes. The PV technology can play a key role in terms of building self-production of electric energy, with a high 'integration' potential into the building envelope. A Building-Integrated Photovoltaic (BIPV) element, by definition, becomes part of the building structure as it is integrated into the envelope, even used in substitution of traditional building components. In this chapter, we define our BIPV concept.

## 1.1 European and Italian Framework

European policy is fostering the use of renewable energies in buildings, setting ambitious goals for the next coming years as foreseen in two strategic directives: the EPBD (Energy Performance Building Directive) 2010/31/EU [3], which states that all new buildings after 2021 will have to be nearly zero-energy and the RES Directive (Renewable Energy Sources) 2009/28/CE [4], that requires minimum levels of RES use in all new buildings.

The European Commission is currently working on a package of measures [5], also updating the above-mentioned Directives, to keep the European Union leadership in clean energy technology and to accelerate clean energy in buildings.

The essential role of the renewable energies in the building sector is also part of the national (Italian) strategy which implemented the RES European Directive in the 'RES national action plan' (June 2010) [6], foreseeing a minimum requirement of electric power from RES in the building sector.

© Springer International Publishing AG, part of Springer Nature 2018
L. Maturi and J. Adami, *Building Integrated Photovoltaic (BIPV)
in Trentino Alto Adige*, Green Energy and Technology,
https://doi.org/10.1007/978-3-319-74116-1_1

In this context, BIPV (Building-Integrated Photovoltaic) systems play an essential role to meet the set EU challenges.

The photovoltaic sector has seen a great development in the past few years, becoming an important part of the energy mix in several EU countries. In Italy in particular, also thanks to the 'Conto Energia' incentive scheme, high overall capacity has been reached. A total cumulative capacity of 18.622 GW is installed and operating in Italy at the end of 2014 [7], i.e. after the conclusion of the 'Conto Energia'. It is extremely interesting to notice that more than half of the total capacity is installed on buildings (Fig. 1.1), in particular, 2.672 GW as BIPV and 7.185 GW as BAPV. As regards to the BIPV plants, it must be noted that about 280 MW plants are related to innovative BIPV plants (as defined under the third, fourth and fifth Conto Energia), while the remaining 2,392 MW of integrated plants include systems on pergolas, greenhouses and shelters (under the second Conto Energia) [7].

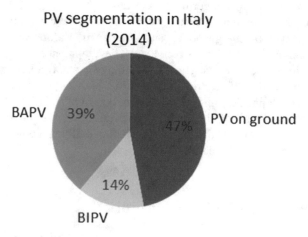

**Fig. 1.1** Segmentation in Italy at the end of 2014 (i.e. after the conclusion of 'Conto Energia') (Eurac research, based on data from [7])

## 1.1.1   Italian Feed-In and BIPV (Edited by GSE)

When Albert Einstein won the Nobel Prize for the photoelectric effect in 1921, he was certainly not thinking about buildings; he could only be disappointed because he was expecting to receive it for his special theory of relativity. After all, the photovoltaic (PV) technology appeared on the roofs of our buildings only 15 years ago, with modules aiming to the sole purpose of producing electricity. It was in the late 90s, and Italy, following the success experienced in Germany, started funding PV plants with the so-called program 'ten thousands photovoltaic roofs'. It gave a special attention to the architectural integration of PV technology granting specific financial incentives.

In July 2005 a feed-in tariff (FiT) system, so-called 'Conto Energia', was introduced, with four successive ministerial decrees. The Regulatory Authority for Electricity and Gas appointed Gestore dei Servizi Energetici, GSE S.p.a., as implementing body incharge for allocating the incentives and setting guidelines related to each FiT law, including BIPV subject.

The second FiT scheme in 2007 had specifically defined the levels of building integration according to three different degrees of implementation (not integrated, partially integrated and integrated), with rising tariffs. Furthermore, an important innovation of the decree was that of taking into account the energy efficiency aspect following the guidelines of the decree 192/05 implementing the European Directive 2002/91/EC on the energy performance of buildings; therefore, a further increase in tariffs was granted for improving the energy efficiency of the annexed building. The category of 'partially integrated plant' plant was designed to take into account the vast majority of the existing building stock, placed in historical towns and making it possible only retrofit installations (so-called BAPV, Building Applied Photovoltaic). Partially integrated systems consisted of PV modules installed both horizontally and inclined over flat roofing or with the same tilt of an underlying sloping roof. With the 'totally integrated plants' category, rewarded with the higher tariff, the decree introduced the concept of substitution of a building component, however without specifying the functions that the PV element should have.

Thus, second Feed-in Law was a first step along the pathway towards architecture; as a matter of fact, the difference between partially and totally integrated plants definition was quite controversial and the introduction of PV modules at a planning stage of the building was uncommon. Therefore, the important PV issue on building functions was missing.

The subsequent decrees (third, fourth and fifth FiT) were issued to bridge the gap, introducing the concept of 'innovative BIPV' through the definition of two principles that are the pillars on which the lintel of PV architecture rests: first, product categories (according to specific construction features) and, second, the installation criteria.

As far as the first one is concerned, a distinction for products coming from two different markets was necessary, because the new design process in which the photovoltaic is a building element involves players of constantly changing sectors. Historical PV producers of standard modules and new entrants from the construction industry attracted by this booming market where to experiment their new PV building products.

The first category includes the so-called special components, standard laminate (frameless) PV modules; for these products, innovation means a patented mounting system specifically designed in order to guarantee waterproofness of the building structure.

Innovative modules are PV building products from construction industry specifically designed and realized for architecture, such as, i.e. flexible PV modules for the substitution of the waterproofing membrane of the entire coverings, thin film layer on rigid support, transparent double glass modules (sometimes with spaced PV cells) designed and installed in order to let the light pass through into the building, and photovoltaic tiles.

Regarding tiles, it must be underlined that installations have to fulfil some requirements on size and the continuity of the whole surface on which the plant is installed; the PV module must have a dimension similar to that of a traditional building element, and in case the PV system is not covering the entire surface of the roof, the plant have to be completed with elements of similar size, guaranteeing the continuity of the whole surface without the need of connecting elements.

Some products showed an intense effort in this new niche market to create new PV custom building products, like, i.e. modules designed in order to reproduce the slate tile composing a plant completed with passive tiles of different sizes specifically inserted to cover the whole surface.

Despite the relevance of the above-mentioned product type eligible for FiT, the main actors of this story are installation criteria and, more in depth, the link between PV and the building. As a matter of fact, the decrees specify that PV modules have to guarantee both the energy production and architectural functions like protection or thermal regulation of the building, moreover introducing the concept of thermal transmittance.

It is quite clear the path where BIPV is going. Photovoltaic technology in buildings is not an electricity generator anymore, but it is also a part of the building structure, replacing traditional elements and guaranteeing their functions. In this sense, installation of PV involves the whole building and not only a surface. PV elements become a structural matter to study, fold, and mold according to building's life, getting part of the architecture, since building envelope performance is carried out by innovative modules and components.

New architectural and urban visions saw the light through five decrees. It is a new dawn for this technology that will not cease with the end of incentive schemes. PV—quoting somehow the renowned words of Louis Kahn, the famous American architect—will always say 'I like Architecture'.

> You say to a brick, 'What do you want, brick?' And brick says to you, 'I like an arch.' And you say to brick, 'Look, I want one, too, but arches are expensive and I can use a concrete lintel.' And then you say: 'What do you think of that, brick?' Brick says: 'I like an arch.'
>
> Louis Kahn

**Disclaimer**

GSE contribution is a simple explanation of the evolution of the feed-in scheme for BIPV. It does not represent any kind of evaluation about eligibility to incentives of PV plants included in the publication.

## 1.2 BIPV Concepts, Definition and Criteria

Several definitions of BIPV have been adopted in the literature and there is no univocal consensus within the PV community and the building sector about BIPV categorization. Within Europe, many countries use different definitions in their feed-in tariffs. For example, in France, the classification of BIPV is strictly related to 'water-tightness' properties, while in Italy the classification is related to several rules and criteria stated in the 'Conto Energia' (Sect. 1.1.1).

At the international level, the recent standard EN 50583-1:2016 'Photovoltaics in buildings' [8], states that photovoltaic modules are considered to be building integrated, if the PV modules form a construction product providing a function as defined in the European Construction Product Regulation CPR 305/2011.

This definition is mainly referred to a technological concept of 'multi-functionality', according to which a BIPV module has to provide additional functions besides the energy production, such as weather protection, thermal insulation, shading, daylighting, noise protection, security.

Anyway, this definition of BIPV strictly related to the technological aspect seems not to be exhaustive enough. Several international projects on BIPV (i.e. PVPS Task 7 [9], IEA Task 41 [10]) has underlined the importance of 'formal/aesthetic' integration, beyond the multifunctionality concept.

Moreover, according to our experience, another aspect is becoming more and more important in relation to BIPV, i.e. the 'energy' integration in the building and district system.

In our opinion thus, the 'I' of the acronym 'BIPV' should stand for 'integration' considering its triple meaning: technology, aesthetic and energy integration (Fig. 1.2).

Despite the existing 'BIPV' definitions, we believe that in order to succeed in the BIPV system design, all three aspects must be considered. In the next paragraphs, a brief explanation of our view on the three 'integration meanings' is provided.

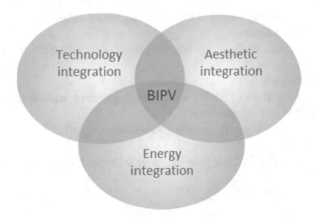

**Fig. 1.2** BIPV integration concepts

### 1.2.1  BIPV: Technology Integration

According to the technological 'multifunctionality' concept, the systems can be divided in two main groups: BAPV (i.e. Building-Added Photovoltaics), and BIPV systems (i.e. Building-Integrated Photovoltaics).

In the first case, PV modules are simply applied on top of the building skin and they are thus commonly considered just as technical devices added to the building, without any specific technical or architectural function. An example of add-on system could be a typical frame-mounted system attached above an existing roof without any architectonical design and not providing any additional functionality to the existing roof.

As for BIPV systems instead, the PV modules are integrated into the envelope constructive system, being an integral part of the building. PV modules, in this case, replace traditional building components and are able to fulfil other functions required by the building envelope (e.g. providing weather protection, heat insulation, sun protection, noise protection, modulation of daylight and security) [11].

### 1.2.2  BIPV: Aesthetic Integration

The aesthetic integration of PV might be described as the capability of the PV solution to define the linguistic/morphological rules governing the design, the structure and the composition of the building's architectural language [12].

In order to define best practice for BIPV aesthetic integration quality, several *architectural criteria* have been defined in the framework of the International Energy agency project IEA-PVPS Task 7 'Photovoltaic power systems in the built environment' [9] and they are hereunder summarized:

- Naturally integrated:
  the PV system is a natural part of the building. Without PV, the building would be lacking something—the PV system completes the building;
- Architecturally pleasing:
  based on a good design, the PV system adds eye-catching features to the architecture;
- Good composition:
  the colour and texture of the PV system is in harmony with the other materials;
- Grid, harmony and composition:
  the sizing of the PV system matches the sizing and grid of the building;
- Contextuality:
  the total image of a building is in harmony with the PV system (e.g. for historic buildings)

- Well engineered:
  the elegance of design details is taken into account. All details are well conceived, the amount of materials is minimized;
- Innovative new design:
  the PV system adds a value to building. The PV is an innovative technology in the field of architecture, asking for innovative, creative, thinking of architects.

After the work carried out by the IEA Task 7 project (in early years of 2000), another more recent IEA project, named IEA-SHC Task 41 'Solar Energy and Architecture' [10], has further developed these concepts by focusing on architectural quality of building-integrated solar energy systems.

Task 41 project defines the 'architectural integration quality' as the result of a controlled and coherent integration of the solar collectors simultaneously from all points of view, functional, constructive and aesthetic [13]. That is when the solar system is integrated in the building envelope (as roof covering, façade cladding, sun shading, balcony fence, etc.), it must properly take over the functions and associated constraints of the envelope elements it is replacing (constructive/functional quality), while preserving the global design quality of the building (aesthetic quality). If the design quality is not preserved (i.e. the system is only constructively/functionally integrated into the building skin without an aesthetic control), the system could be only be defined as 'building integrated' rather than 'architectonically integrated' [13].

## 1.2.3   BIPV: Energy Integration

The energy integration refers to the capability of a PV system to interact with the building and district energy system in order to maximize the local use of the produced electricity.

In our opinion, the energy integration will become more and more important to cope with new ways to conceive buildings and their energy provision. In fact, also thanks to the EU policy oriented to promote the NZEB (nearly zero-energy buildings) concept and RES (renewable energy sources) exploitation [3, 4], buildings are becoming more than just stand-alone units using energy from the grid. They are becoming energy partners in the energy system as micro energy hubs consuming, producing, storing and supplying energy, thus transforming the EU energy market, shifting from centralised, fossil-fuel based, national systems towards a decentralised, renewable, interconnected and variable system.

# References

1. OECD/IEA and AFD (2008) Promoting energy efficiency investments
2. IEA (2009) Task 40—towards net zero energy solar buildings
3. European Commission (2010) Directive 2010/31/EU of the European Parliament and the Council on the energy performance of buildings (EPBD). 19 May 2010
4. European Parliament and European Council (2009) Directive 2009/28/CE on renewable energy sources (RES). 23 Apr 2009
5. European Commission (2016) Clean energy for all Europeans
6. Ministero dello sviluppo economico (2010) Piano di azione nazionale per le energie rinnovabili (direttiva 2009/28/CE)
7. IEA (2014) Task 1—national survey report of PV power applications in Italy 2014
8. CENELEC (2016) EN 50583—photovoltaics in buildings
9. IEA (2001) Task 7 of the PV power systems program—achievements and outlook
10. IEA (2008) Task 41—solar energy and architecture-annex plan
11. Maturi L (2013) Building skin as energy supply: prototype development of a wooden prefabricated BIPV wall
12. Swiss BiPV Competence Centre, "BIPV presentation"
13. IEA (2012) Task 41 A.2—solar energy systems in architecture—integration criteria and guidelines

# Chapter 2
# BIPV Architectural Systems

**Abstract** PV modules can be integrated into different parts of the building envelope, creating specific architectural systems. In this chapter different categories of BIPV systems are explored, including roof, façade and external devices. These categories require different technological ways of using PV in the envelope, which lead to different choices of the PV components and materials.

## 2.1 BIPV Integration Possibilities into the Envelope

PV modules can be integrated into different parts of the building fabric and it is possible to group them into three macro-categories: facade systems (which include curtain wall products, spandrel panels, glazings, etc.), roof systems (which include tiles, shingles, standing seam products, skylights, etc.) and external devices (e.g. parapets, external shading devices).

Between these categories, mixed configurations are also possible, as in the case of many contemporary architectures, where a clear distinction between roof and façade does not exist anymore.

In the following, a brief overview of the above-mentioned categories is presented, according to the review performed by IEA Task 41 experts [1]. Some of the presented products might not be on the market anymore, since the BIPV products market is changing continuously, but it is anyway interesting to show examples of various possible solutions. Several databases collecting information on existing BIPV products on the market are available online, e.g. [2, 3, 4], and might be useful to look for new products.

### 2.1.1 Roof Integration

According to different levels of 'technology integration', a PV component can be added on the roof, it can substitute the external layer of the roof system (i.e. PV as a

L. Maturi and J. Adami, *Building Integrated Photovoltaic (BIPV)
in Trentino Alto Adige*, Green Energy and Technology,
https://doi.org/10.1007/978-3-319-74116-1_2

cladding), or it can also substitute the whole technological sandwich (i.e. semi-transparent glass–glass modules as skylights). Depending on the layers in the PV component substitutes, it has to meet different requirements that influence the choice of the most suitable PV component.

**Opaque Tilted Roof**

Building-added PV systems are very commonly used on tilted roofs especially in case of retrofit systems. Using this solution, there is a need for an additional mounting system and in most cases of the reinforcement of the roof structure due to the additional loads.

In some cases, the building-added PV systems on the roof have been highly criticized for their aesthetics that urged the market to provide building-integrated products replacing all types of traditional roof claddings, especially for special contexts, such as historic centres.

Thus, several products were developed both with crystalline and thin-film technologies for roof tiles, shingles and slates that aesthetically match with common roof products (Fig. 2.1).

**Fig. 2.1** Tilted roof systems. Left: solar tile © Tesla Solar. Centre: solar shingles © Panotron. Right: solar slates © Sunstyle

Several metal roof system manufacturers (standing seam, click-roll-cap, corrugated sheets) developed their own BIPV products with the integration of thin-film solar laminates. Some of these products are even conceived as prefabricated roofing systems which incorporate thermal insulation [1].

**Opaque Flat Roof**

In the case of flat opaque roofs, the most commonly used systems are added systems with rack supporting standard glass-Tedlar modules, or specific tilted rack systems with thin-film laminates (Fig. 2.2, left).

**Fig. 2.2** Flat roof systems. Left: special rack system for flexible laminate on stainless steel substrate © Unisolar. Centre: Powerply monocrystalline module with plastic substrate © Lumeta Solar. Right: flexible laminate on metal substrate © Kalzip

There is also a possibility to use crystalline modules with plastic substrates allowing a seamless integration on the roof with an adhesive backing (Fig. 2.2, centre). Thin-film technologies also offer different flexible laminates, with plastic or stainless steel substrates, that can be easily mounted on flat roofs (Fig. 2.2, right). Another common trend for a flat roof is using the waterproof membrane as a support on which flexible thin-film laminates are glued, providing a simple and economic integration possibility [1].

It is important that the material used as a module substrate is in compliance with the fire-fighters requirements.

**Semi-transparent Roof**
The PV system can also become the complete roof covering, fulfilling all its functions. Most commonly semi-transparent crystalline or thin-film panels are used in skylights (Fig. 2.3). These solutions provide controlled daylighting for the interior, while simultaneously generating electricity [5].

**Fig. 2.3** Semi-transparent skylights. Left: residential building, Austria: semi-transparent modules with crystalline cells © Ertex Solar. Centre: residential building, California: semi-transparent modules with a stripe pattern of crystalline cells © Atlantis Energy Systems. Right: Valladolid University, Spain: semi-transparent thin-film modules © Onyx Solar

## 2.1.2  Façade Integration

According to different levels of 'technology integration' described in the previous chapter, a PV component can substitute the external layer of the facade (i.e. PV as a cladding of a cold facade), or it can substitute the whole façade system (i.e. curtain walls—opaque or translucent) [6, 7, 8].

In the following, a general overview of the way PV can be used in facades will be presented, according to the review performed by IEA Task 41 experts [1].

In the following, the opaque facades are divided into two macro-categories: 'cold' (if the weather protection is provided by a back-ventilated layer) and 'warm' facades (if the weather protection is provided by a not back-ventilated layer) [9].

**Opaque Cold Façade**
In opaque cold facades, the PV panel is usually used as a cladding element. The cladding (PV) is hanged on a substructure anchored to the load-bearing wall.

In these cases, the PV performance might take advantage of its back-ventilation, keeping a lower PV operating temperature. Recently, innovative PV modules have been developed with the aim to fully cover the PV cells making the PV module completely 'mimetic' with a 'standard material' appearance (Fig. 2.4, centre). Typically, they are based on highly efficient solar cells (e.g. Si-PERC), covered with a special layer (e.g. printed on the glass).

**Fig. 2.4** Facade cladding solutions. Left: Solara Hause, Switzerland: monocrystalline modules © MGT-esys. Centre: monocrystalline nanotechnology-based modules © Issol & Solaxess. Right: Rosenheim Institut für Fenstertechnik, Germany: thin-film modules © Schüco

## Opaque Warm Façade
Curtain wall systems (Fig. 2.5) also offer opportunities for PV integration. In this case, the PV might be integrated either on the transparent part (semi-transparent PV) or on the opaque part (spandrel). For the latter, depending on the used PV technology, it might be important to foresee a PV back-ventilation.

**Fig. 2.5** Warm façade solutions. Left: Zara Fashion Store of Cologne, Germany: opaque monocrystalline cells combined with transparent glazing in post-beam curtain wall structure © Solon. Centre: GDR Headquarter, Spain: thin-film modules © Onyx Solar. Right: Zwolle multi-storey parking, the Netherlands: thin-film modules © Schott Solar

## Semi-transparent Façade
Semi-transparent PV modules can be integrated into semi-transparent parts of the façade (Fig. 2.6). With crystalline cells, the distance among each cell inside the module can be freely defined, controlling transparency and aesthetical effect (Fig. 2.6, left and centre).

Also, a single crystalline cell can be semi-transparent (due to grooved holes in the cell), but this solution is rarely used for its costs.

With thin-film modules, transparency is created by additional grooves among the cell strips, creating a finely checked pattern that gives the thin-film modules a neutrally homogeneous transparency (Fig. 2.6, right) [1].

**Fig. 2.6** Semi-transparent façade. Left: crystalline glass modules © Issol. Centre: crystalline glass modules © EnergyGlass. Right: semi-transparent thin-film modules © Sanyo

## 2.1.3  External Devices

Photovoltaics can also be used as external devices on the building skin, integrated as shading devices, louvres or canopies, movable shutters, parapets, balconies and in general as architectonical elements added to the building (Fig. 2.7). Shading systems are the most commonly used. Semi-transparent glass–glass components with semi-transparent crystalline or thin-film are very often used. On balconies, glass semi-transparent modules made of security glass are mostly used. Opaque systems are also widely used.

**Fig. 2.7** External devices. Left: Le Albere district, Italy: semi-transparent modules integrated into louvres © FAR Systems. Centre: semi-transparent modules integrated into a canopy © Ertex Solar. Right: Solarsiedlung Hintere Laugeten, Switzerland: thin-film module

## References

1. IEA (2012) Task 41 A.2—solar energy systems in architecture—integration criteria and guidelines
2. "Institute for Renewable energy Eurac" [Online]. Available: http://www.eurac.edu/it/research/technologies/renewableenergy/Pages/default.aspx
3. SUPSI, "BIPV building integrated photovoltaic" [Online]. Available: http://www.bipv.ch/index.php/en/
4. IEA, "Database of innovative solar products for building integration" [Online]. Available: http://solarintegrationsolutions.org/
5. Frontini F, Kuhn TE (2010) A new angle-selective, see-through BIPV façade for solar control

6. Maturi L (2013) Building skin as energy supply: prototype development of a wooden prefabricated BIPV wall
7. Quesada G, Rousse D, Dutil Y, Badache M, Hallé S (2012) A comprehensive review of solar facades. Transparent and translucent solar facade
8. Yun GY, McEvoy M, Steemers K (2007) Design and overall energy performance of a ventilated photovoltaic façade
9. Knaack U, Klein T, Bilow M, Auer T (2010) Fassaden: Prinzipien der Konstruktion, Birkhäuser

# Chapter 3
# Case Studies

**Abstract** This chapter shows a collection of 'ordinary BIPV high-quality examples' realized in the 'Trentino Alto Adige' region. The best and most representative case studies were selected out of more than 40 collected buildings. They include several kinds of integration typologies representing both private and public sector, giving an overview of different approaches to the BIPV matter, especially regarding the decision-making related to economic issues. A detailed description of each case study is provided in order to evaluate the BIPV projects from different points of view, starting from the design process phases.

## 3.1 Case Studies Selection Methodology

In order to collect the best and most representative case studies present in the 'Trentino Alto Adige' region, a 'call for case studies' was launched by contacting most of the engineers, architects and professionals of the region through several channels (Eurac contact list, Engineers and Architect associations).

Out of more than 40 collected cases, the best ones were selected through an internal workshop involving 19 Eurac collaborators, forming a 'selection group' (Fig. 3.1).

**Fig. 3.1** Pictures of the workshop for the selection process

© Springer International Publishing AG, part of Springer Nature 2018
L. Maturi and J. Adami, *Building Integrated Photovoltaic (BIPV)
in Trentino Alto Adige*, Green Energy and Technology,
https://doi.org/10.1007/978-3-319-74116-1_3

The selection was based on two main criteria, i.e. 'overall impression/global approach' and 'lesson learnt'. The collaborators were asked to evaluate each case study by considering the following aspects:

- Overall impression/global approach, considering:
  - the way PV was integrated into the whole concept of the building and contributed to/preserve architectural quality;
  - the whole building concept;
  - surface texture, the composition of visible materials, correlations of colours, details (especially successful new designs)/joints/fixing.

- Lesson learnt, considering the following questions:
  - Is it an interesting experiment of PV integration beyond the achieved results?
  - Is there something to learn about it?
  - Are there innovative architectural concepts, original solutions, original applications?

The cases getting a score higher than 20 marks were selected and included in this publication, meaning that the great majority of the 'selection group' found the project interesting and inspirational for architects.

In this way, 19 best case studies were selected. They are spread out across the Trentino Alto Adige region (Fig. 3.5).

For each of the selected case studies, a questionnaire was prepared to collect information from at least one of the involved stakeholders (e.g. architect, owner, PV system electrical designer, installer, general contractor).

## 3.2   Case Studies Overview

The case studies were categorized according to the EN standard 50583:2016 'Photovoltaics in buildings' [1], as shown in Fig. 3.2 and in Table 3.1.

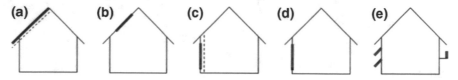

**Fig. 3.2** 'Photovoltaics in buildings' categories according to [1]

**Table 3.1** Case studies characterization according to their BIPV architectural system

|    | BIPV case study | Architectural system | Standard category | Ownership | Location |
|----|-----------------|----------------------|-------------------|-----------|----------|
| 1  | Ütia da Ju | Semi-transparent roof | B | Private | San Martino di Badia (BZ) |
| 2  | Milland Church | Opaque tilted roof | A | Public | Bressanone (BZ) |
| 3  | Colterenzio Winery | Semi-transparent roof | B | Private | Appiano (BZ) |
| 4  | Single-family house | External semi-transparent device (parapet) | E | Private | Lasa (BZ) |
| 5  | Farm building | Semi-transparent roof | B | Private | San Genesio Atesino (BZ) |
| 6  | Enzian Office | Opaque cold façade, semi-transparent façade | C, D | Private | Bolzano (BZ) |
| 7  | La Pedevilla Chalet | Opaque tilted roof | A | Private | Marebbe (BZ) |
| 8  | Hafner Tower | External opaque device (shading) | E | Private | Bolzano (BZ) |
| 9  | Multi-family house | Opaque tilted roof | A | Private | Appiano (BZ) |
| 10 | Autobrennero headquarter | Opaque cold façade | C | Private | Trento (TN) |
| 11 | Ex-Post | Opaque cold façade | C | Private | Bolzano (BZ) |
| 12 | Le Albere district | External semi-transparent device | E | Public | Trento (TN) |
| 13 | Maso Lampele | Opaque tilted roof | A | Private | Novacella di Varna (BZ) |
| 14 | District heating plant | Opaque cold façade, semi-transparent façade | C, D | Public | Laces (BZ) |
| 15 | Smart Lab | Opaque cold façade | C | Public | Rovereto (TN) |
| 16 | Cable car station | Semi-transparent roof-façade | B, D | Private | Naturno (BZ) |
| 17 | Chamber of Commerce | Opaque cold façade | C | Public | Bolzano (BZ) |
| 18 | Castaneum Center | Opaque tilted roof | A | Public | Velturno (BZ) |

The categories schematized in Fig. 3.2 can be described as follows:

- Category A: Sloped, roof-integrated, not accessible from within the building
- Category B: Sloped, roof-integrated, accessible from within the building
- Category C: Non-sloped (vertically) mounted not accessible from within the building
- Category D: Non-sloped (vertically) mounted accessible from within the building
- Category E: Externally integrated, accessible or not accessible from within the building

Among the selected case studies, examples of different type of building typologies are presented, including office, residential, agricultural, industrial, community, religious, commercial and transportation buildings.

Several architectural integrations are shown, including opaque and semi-transparent roof, warm, cold and double skin facades as well as external devices such as parapets and solar shading elements. The most predominant are façade and roof systems (Fig. 3.3).

Different kind of ownerships (private and public) are also represented, giving an overview of different approaches to the BIPV matter, especially regarding the decision-making and the design and installation process.

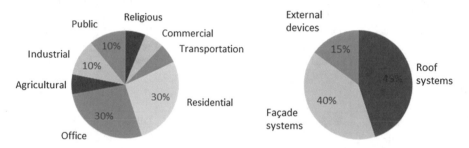

**Fig. 3.3** Building typologies and architectural integration types presented in the case studies

Regarding the PV modules, the crystalline technology is the mostly used, being applied in around 80% of the analysed case studies. Most of the systems are made using standard modules (only 30% are made with custom modules) showing that in many cases appealing BIPV systems can be realized without the need of customization (Figs. 3.4).

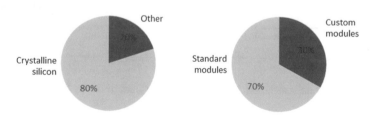

**Fig. 3.4** PV technologies and module types presented in the case studies

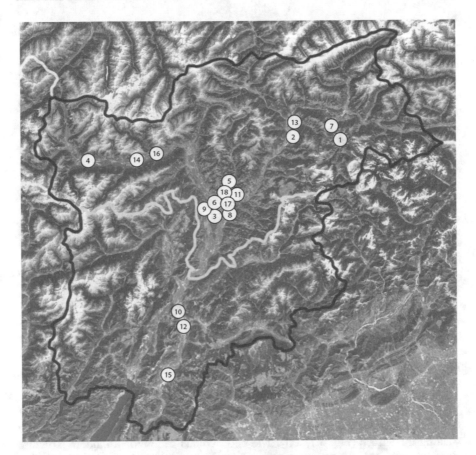

**Fig. 3.5** Localization of the selected case studies, Trentino Alto Adige map

## 3.2.1 Ütia da Ju, San Martino di Badia (BZ)

| PROJECT DATA | |
| --- | --- |
| **PROJECT TYPE** | Retrofit |
| **BUILDING FUNCTION** | Commercial |
| **INTEGRATION SYSTEM** | Semi-transparent roof |
| **LOCATION** | Strada Ju 43, San Martino di Badia (BZ) |

### BUILDING/SYSTEM INTEGRATION

#### Aesthetic integration

The PV system is integrated into the shading canopy applied to a tourist accommodation, hosting a restaurant, placed in Piz de Plaies (Fig. 3.6). The solar canopy has a distinctive and original spiral structure exposed on a steep wooded slope of

**Fig. 3.6** Ütia da Ju BIPV system hanging out from the building (Augustin Clement)

Val Badia, at an altitude of 1622 m. It consists of semi-transparent trapezoidal PV modules running around the building roof with a uniform inclination and partially shading the terrace (Fig. 3.7, left). The photovoltaic canopy is a modern structure harmoniously and elegantly integrated into the natural environment of a region bound to the traditional Ladin culture and language (Fig. 3.7, right).

**Fig. 3.7** Left: the building is embedded in the natural surroundings of Piz de Plaies (Augustin Clement). Right: solar canopy exposed on the wooded slope (Augustin Clement)

**Energy integration**

The BIPV system BIPV system has been designed to have an annual electricity production of about 7,800 kWh to cover almost the whole building electricity demand. It feeds additional energy into the grid (Augustin Clement). Thanks to the module's semi-transparency, the photovoltaic roofing system allows part of the sunlight to enter through the structure, contributing to the heating of the canopy, which is an enclosed space.

**Technology integration**

The 40 semi-transparent modules (Ertex VSG 140-211) were customized in order to match the building shape. Moreover, they had to get the specific texture and mechanical resistance as UNI7696 (Ertex Solartechnik GmbH). The modules are composed of laminated safety glass, encapsulating polycrystalline cells 2 cm spaced out (Fig. 3.8, left). They are supported by a timber load-bearing structure, specially designed according to the modules' shape. The structure beams hide the module's fixing system and the cables from people staying below (Fig. 3.8, right). All the electric connections are made of special UV-resistant cables. They are protected with waterproof ducts (Fig. 3.9, left).

**Fig. 3.8** Left: construction phase, the frameless modules are easily mounted (Augustin Clement). Right: timber structure hiding the modules fixing system (Augustin Clement)

**Fig. 3.9** Left: view from above of the BIPV glass roof: the visible ducts protect the electric connections (Augustin Clement). Right: BIPV shading effect (Augustin Clement)

## DECISION-MAKING

After discovering some interesting examples of BIPV installations at an Austrian exhibition, the restaurant owner decided to apply the photovoltaic technology. He wanted to integrate it into the building in order to create an installation visible to all the visitors, aesthetically attractive and functionally useful also as a shading device. Even though the building is placed into a mountain environment, introducing an innovative element as a BIPV system did not raise any problem getting the planning permission from the public administration (Augustin Clement).

## PROCESS

The PV system was integrated in 2009, two years after the construction of the building. The client, the restaurant owner Augustin Clement, entrusted Electro Clara Sas as main responsible for the BIPV system. Once carried out an energy assessment (online PVGIS) (Ertex Solartechnik GmbH), the company provided the technical design and the installation of the BIPV system. Ertex Solartechnik GmbH was chosen as modules supplier and Prada Holzbau Srl as canopy constructor.

## LESSON LEARNT

Ütia da Ju represents an exemplary case of a retrofit, where a complex structure bearing the integrated PV plant was applied after the building construction. The particular configuration of the solar canopy implies a great attention to details and specific construction solutions. The BIPV system has to guarantee the same functions as the traditional roofing system, e.g. water tightness, resistance to snow loading, to atmospheric agents, to fire, etc. It required an accurate design process which also took into account the aesthetic viewpoint. Finally, it resulted to be an expressive architectural element thanks to the module's aesthetic appearance and the shading effect under the structure. The solar canopy covers an area which was equipped to accommodate the restaurant customers that can appreciate the attractive atmosphere (Fig. 3.9, right).

| BIPV SYSTEM DATA | |
|---|---|
| PV MODULES | Custom made |
| SOLAR TECHNOLOGY | Polycrystalline silicon |
| NOMINAL POWER | 7.05 kWp |
| SYSTEM SIZE | 100 m$^2$ |
| MODULE SIZE | / |
| ORIENTATION | From 75° West to 45° East |
| TILT | 10° |

| BIPV SYSTEM COSTS | |
|---|---|
| TOTAL COST (€) | 36,200 |
| €/m$^2$ | 362 |
| €/kWp | 5,135 |

| PRODUCER DATA | |
|---|---|
| **PRODUCER** | Ertex Solartechnik GmbH |
| **ADDRESS** | Mitterhofer Strasse 4, Amstetten (A) |
| **CONTACT** | info@ertex-solar.at |
| **WEB** | www.ertex-solar.at |

## 3.2.2   Milland Church, Bressanone (BZ)

| PROJECT DATA | |
|---|---|
| **PROJECT TYPE** | Retrofit |
| **BUILDING FUNCTION** | Religious |
| **INTEGRATION SYSTEM** | Opaque tilted roof |
| **LOCATION** | Via Campill, Bressanone (BZ) |

### BUILDING/SYSTEM INTEGRATION

#### Aesthetic integration

The PV system is a pitched roofing solution integrated into a metal sheets roof (Fig. 3.10). It represents an interesting example of retrofit of a church built in 1984–1985, where the original building shape, composition and main colours are

**Fig. 3.10** Milland Church BIPV system installed on the SE-facing slope (Eurac research)

respected (Fig. 3.11, left). The roofing system is divided into six triangular parts. The PV plant, installed on the south-west facing part, is made of dark modules with black rear side base in order to keep homogeneity in surface and colours (Fig. 3.11, right).

**Fig. 3.11** Left: view of the original metal roofing surface, not so different from the PV modules (Eurac research). Right: construction phase, the church metal roof and the BIPV mounting system are still visible (Eurac research)

### Energy integration
The BIPV system annually produces an electrical output of around 22,000 kWh (as monitored in 2011 by Eurac research). It supplies most of electrical energy need of the church, the parish centre and the rectory (Don Vijo Luigi Alois Pitscheider).

### Technology integration
87 PV modules (SunPower Co., SPR-220 BLK) are integrated 14.5 cm far from the outside roof layer with a gap of 2.5 cm among the module arrays. The air gap allows natural ventilation of the PV modules and thus slightly reduces the power losses caused by the increased operative temperature (Fig. 3.12, left). The chosen module typology eliminates standard metal gridlines, since metal contacts are placed on the back of the solar cell, out of sight. The fixing system is made of metal rails and clips (Fig. 3.12, right).

**Fig. 3.12** Left: modules are naturally back ventilated (Eurac research). Right: detailed view of the BIPV fixing structure (Eurac research)

## DECISION-MAKING

The main reason that leads the priest Don Vijo Luigi Alois Pitscheider to strongly want the integration of the photovoltaic system into the Milland Church roof was making a practical contribution to promote the use of renewable energy, serving an example for the community. Using one of the most sun-exposed roof parts to produce electric energy was also an opportunity to enhance the modern church features. Once the feasibility and cost effectiveness of the intervention were assessed, the priest succeeded in overcoming an initial disagreement with the local government (Don Vijo Luigi Alois Pitscheider).

## PROCESS

The first step was to obtain the religious institution's permission to install such an innovative technology on a holy building. Once the priest has defined the PV plant configuration thanks to the technical advice of the team Von Lutz and the architectural design of Arch. Claudio Paternoster, Elettropiemme Srl was entrusted with the BIPV system electric design and installation. At the suggestion of Elettropiemme Srl, the American SunPower Corporation was chosen, having a product with the right colour and a good energy performance (Don Vijo Luigi Alois Pitscheider). The works were concluded in 2008.

## LESSON LEARNT

The chosen PV technology, based on back contact technology, combines an esthetical appeal (homogeneous black appearance) with the energy performance (the PV cells are entirely exposed to solar radiation without any covering due to standard front contacts). Fake modules play an important role in the finishing of the BIPV system. In order to respect the roof geometry, several fake modules have been installed around the perimeter of the PV system. First fake modules which have been installed resulted to be too reflective and did not match very well with the active PV modules in the considered operative conditions (Fig. 3.13, left). They have thus been removed and replaced with less reflective ones in order to provide a more homogeneous and coherent aspect (Fig. 3.13, right). This experience underlines the importance of finishing details, which can make the difference to reach high quality in BIPV systems.

**Fig. 3.13** Left: first fake modules installed, too reflective (Eurac research). Right: final fake modules installed, matching the BIPV plant surface (Eurac research)

| BIPV SYSTEM DATA | |
|---|---|
| PV MODULES | Standard |
| SOLAR TECHNOLOGY | Monocrystalline silicon |
| NOMINAL POWER | 19 kWp |
| SYSTEM SIZE | 107 m$^2$ |
| MODULE SIZE | 1,559 × 798 mm |
| ORIENTATION | South-west |
| TILT | 35° |

| BIPV SYSTEM COSTS | |
|---|---|
| TOTAL COST (€) | 140,000 |
| €/m$^2$ | 1,308 |
| €/kWp | 7,315 |

| PRODUCER DATA | |
|---|---|
| PRODUCER | SunPower Corporation |
| ADDRESS | Rio Robles San Jose 77, California |
| CONTACT | / |
| WEB | www.sunpower.com |

### 3.2.3  Colterenzio Winery, Appiano (BZ)

| PROJECT DATA | |
|---|---|
| PROJECT TYPE | Retrofit |
| BUILDING FUNCTION | Industrial |
| INTEGRATION SYSTEM | Semi-transparent roof |
| LOCATION | Str. Vino 8, Appiano (BZ) |

## BUILDING/SYSTEM INTEGRATION

### Aesthetic integration

The Colterenzio Winery headquarter comes from the refurbishment of existing structures from the 1970s. The old buildings have undergone a major restoration that integrated the tradition of the oak wood with the innovation of modern materials and technologies (Fig. 3.15, left). The BIPV system is part of the translucent glass roof of a steel canopy placed within the building's complex (Fig. 3.15, right). It is made of semi-transparent modules that provide an aesthetically attractive shading effect (Fig. 3.14).

**Fig. 3.14** Colterenzio winery BIPV system (Günter Richard Wett)

**Fig. 3.15** Left: integration of different materials and structures (Günter Richard Wett). Right: view of the glass photovoltaic roof and the black aluminium grill placed below (Günter Richard Wett)

### Energy integration

The integrated photovoltaic system was calculated to have an annual production of 28,300 kWh (Eng. Klaus Fleischmann). It contributes, together with the other PV system installed on the headquarter buildings, to 55% of the winery electricity consumption, mainly due to the machineries and the air conditioning system of the wine cellar. The most of the generated electricity is self-consumed by the buildings. The use of the photovoltaic technology is added to a combined technique of solar panels and efficient capture of the heat, able to provide 70% of all hot water required (Colterenzio Winery).

**Technology integration**

The 184 integrated PV modules Solarwatt (M140-36 GEG LK XL) are standard semi-transparent panels made of 36 monocrystalline cells placed between glazed laminates (Fig. 3.16, left). They are mounted as a simple glass into the canopy roofing system and naturally ventilated. The bearing structure is made of steel profiles that guarantee the roof water tightness (Fig. 3.16, right).

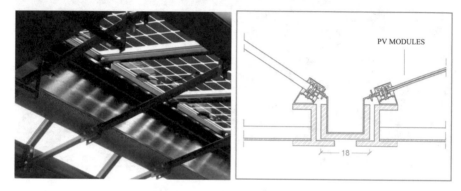

**Fig. 3.16** Left: detailed view of the BIPV plant bearing structure (Colterenzio Winery). Right: technical detail of the modules fixing system, re-drawn by Eurac research (Eng. Klaus Fleischmann)

**DECISION-MAKING**

Since 2009, the Colterenzio Winery headquarter has been under renovation to be harmonized and steered towards environmental sustainability. Energy saving, costs reduction and an environmentally friendly high-quality wine production are the main goals that led the company to equip all buildings with photovoltaic and solar panels. A semi-transparent BIPV plant was integrated on the canopy in order to achieve a high aesthetic quality. It did not require substantial changes to the original architectural appearance. No special requirements in relation to the building permit were needed (Colterenzio Winery).

**PROCESS**

The design process involved the Colterenzio Winery chairman Maximilian Niedermayr and the project teams, including the bergmeisterwolf architecture firm which gave to the old winery a new appearance (together with the architecture collaborators Roland Decarli and Jürgen Prosch), and the Fleischmann & Jansen engineering firm, which provided the structural design and an energy and economic assessment of the BIPV plant. Solarwatt AG provided the photovoltaic modules.

The installation was entrusted to Obrist Gmbh and Elektro Ebner Gmbh. It was concluded in 2014.

**LESSON LEARNT**

Achieving a high aesthetic quality was one of the main design purposes, together with the energy performance. Thanks to the BIPV shading effect, the space covered by the solar canopy takes on an attractive appearance, a composition of light and shadow changing throughout the day (Figs. 3.17, left and right). This allows an area usually used for the grapes storage to be used also for different purposes. The canopy is sometimes a place used as a venue for events (bergmeisterwolf). This case study showed how the BIPV technology can be used in order to improve the potentialities of a construction. All its functions should be analysed from the beginning of the design process to maximize the PV integration benefits. So, all the stakeholders (owner, designers, consultants, technics, etc.) should be involved to work together (bergmeisterwolf).

**Fig. 3.17** Left: BIPV shading effect (Günter Richard Wett). Right: the solar canopy takes on an aesthetically pleasing appearance (Günter Richard Wett)

| BIPV SYSTEM DATA | |
| --- | --- |
| PV MODULES | Standard |
| SOLAR TECHNOLOGY | Monocrystalline silicon |
| NOMINAL POWER | 27.7 kWp |
| SYSTEM SIZE | 236 m$^2$ |
| MODULE SIZE | 1,600 × 800 mm |
| ORIENTATION | South |
| TILT | 20° |

| BIPV SYSTEM COSTS | |
| --- | --- |
| TOTAL COST (€) | 138,775 |
| €/m$^2$ | 589 |
| €/kWp | 5,010 |

| PRODUCER DATA | |
| --- | --- |
| PRODUCER | Solarwatt AG |
| ADDRESS | Maria Reiche Straße 2A, Dresda (D) |
| CONTACT | info@solarwatt.net |
| WEB | www.solarwatt.net |

## 3.2.4   Single-Family House, Lasa (BZ)

| PROJECT DATA | |
| --- | --- |
| PROJECT TYPE | New construction |
| BUILDING FUNCTION | Residential |
| INTEGRATION SYSTEM | External semi-transparent device (parapet) |
| LOCATION | Via Venosta 70/a, Lasa (BZ) |

### BUILDING/SYSTEM INTEGRATION

#### Aesthetic integration

The PV system is integrated into a 2-storey residential building located in a small village of Val Venosta, along the Adige River (Fig. 3.18). It consists of

**Fig. 3.18** BIPV system as central component of the main building façade (building owner)

semi-transparent glass modules installed in the glazed balconies railings on the first level (Fig. 3.19, left). The modules represent a barrier that protects the large windows characterizing the main building façade, without blocking the mountain landscape view from inside (Fig. 3.19, right). Their pattern highlights the building's horizontal development. Due to the refined design, the BIPV system combines the energy production functionality with an aesthetically pleasing aspect.

**Fig. 3.19** Left: view of the two photovoltaic railings (building owner). Right: the semi-transparent railing allows to enjoy the landscape from inside (building owner)

**Energy integration**
The BIPV plant was designed to provide a yearly energy of around 800 kWh. Its electricity output, together with the production of additional PV modules located on the roof (1 kWp), supplies the energy demand of a connected PV-Heater (Refusol), which is used to heat up tap water with a heating rod in the house's hot water tank. The two PV plants form a stand-alone system which is able to cover the whole building's thermal energy need (building owner).

**Technology integration**
The BIPV plant is made from 6 frameless modules (EGM 84-90 ST), which are assembled using laminated safety glass (10 + 10 mm). The PV cells between the glass layers are spaced out leaving gaps of 2–5 cm, thus making the modules semi-transparent (37–38%) (Fig. 3.20, left). The modules are connected to inner bypass diodes, which do not require the modules to be divided into sub-modules. Two junction boxes are placed at the bottom of each glass panel (Fig. 3.21, left). The PV mounting system (Q railing easy glass slim) does not require holes because the laminated glass is wedged into a 120 mm metal rail all along the balcony which also guarantees the water drainage (Fig. 3.20, right).

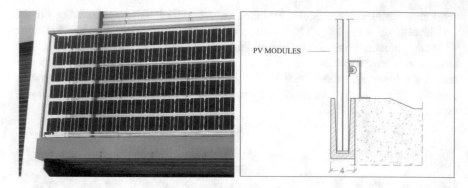

**Fig. 3.20** Left: the crystalline cells partially protect the large windows from outside view (building owner). Right: technical detail of the 'Q railing' mounting system, re-drawn by Eurac research (building owner)

## DECISION-MAKING

The owner decided to integrate photovoltaic modules into the balcony's railings when the building construction was almost completed. Primarily, the PV plant is a useful solution to supply the boiler energy demand, previously supplied by a pellet stove. Second, the owner wanted to use a semi-transparent shading device to partially cover the view into the large windows, initially thinking about a satin or serigraphic glass solution. The final BIPV solution was found visiting a PV products exhibition, where he compared different solar glass solutions and found the best one (building owner).

## PROCESS

In 2007, the chartered building surveyor Coletti Renato was commissioned by the owner to design the house. The construction works took about 4 years. After comparing different quotes of PV manufacturers, EnergyGlass Srl was chosen for the modules supply and the BIPV system design. EnergyGlass Srl also managed the bureaucracy around receiving government incentives. The owner installed the system himself with the help of specialists. Later, additional PV modules were applied on the building roof in order to supply the total boiler energy demand (building owner).

## LESSON LEARNT

The building owner carried out a detailed evaluation before deciding to integrate the photovoltaic technology in the glazed parapet. He wanted something that could partially cover the windows, so he also considered to install satin or serigraphic glass. An economic assessment revealed that the glazed PV could be quite competitive with the glass. Aesthetically, quite the same striped texture could be produced (Fig. 3.21, right). So, the photovoltaic option has been preferred (building owner). The low amount of energy production and the lack of a suitable storage solution on market, in 2012, led the owner to connect the photovoltaic plant to the PV heater exploiting in a different way the generated electricity. The current innovation level reached on the photovoltaic market allowed him to re-think other possible solutions, as using an inverter with integrated energy storage (building owner). This confirms that the energy integration aspect is becoming more and more important in BIPV.

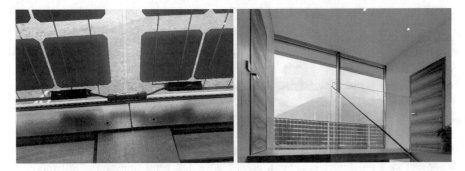

**Fig. 3.21** Left: detailed view of the modules cables junction (building owner). Right: view of the photovoltaic railing from inside (building owner)

| BIPV SYSTEM DATA | |
| --- | --- |
| PV MODULES | Custom made |
| SOLAR TECHNOLOGY | Monocrystalline silicon |
| NOMINAL POWER | 1.3 kWp |
| SYSTEM SIZE | 13 m$^2$ |
| MODULE SIZE | 1,120 × 1,905 mm, 1,120 × 2,005 mm |
| ORIENTATION | South |
| TILT | 90° |

| BIPV SYSTEM COSTS | |
| --- | --- |
| TOTAL COST (€) | 5,992 |
| €/m$^2$ | 461 |
| €/kWp | 4,609 |

| PRODUCER DATA | |
| --- | --- |
| PRODUCER | EnergyGlass Srl |
| ADDRESS | Via Domea 79, Cantù (CO) |
| CONTACT | / |
| WEB | www.energyglass.eu |

## 3.2.5   Farm Building, San Genesio (BZ)

| PROJECT DATA | |
|---|---|
| PROJECT TYPE | Retrofit |
| BUILDING FUNCTION | Agricultural |
| INTEGRATION SYSTEM | Semi-transparent roof |
| LOCATION | Brunner Avigna 1, San Genesio (BZ) |

### BUILDING/SYSTEM INTEGRATION

#### Aesthetic integration

The building, an old construction, was retrofitted through integrating a PV system into the building roof (Fig. 3.22). It is a traditional 2-storyes farmhouse, with a barn upstairs and stalls downstairs, currently used as farm equipment storage (Fig. 3.23, left). The building is located in a little village, high above the valley entrance of the Sarentino Valley. It is embedded in the charming scenery of the Salto high plateau, far away from the main traffic lines. The BIPV system is modern technology surrounded by a natural landscape characterized by meadows, larch trees and traditional buildings (Fig. 3.23, right). The semi-transparent PV modules create an interesting light and shadow pattern inside.

**Fig. 3.22**  Farm building BIPV system (Ing. Studio Blasbichler Co. Ltd.)

**Fig. 3.23** Left: the building embedded in the mountain surrounding of the Salto high plateau (Ing. Studio Blasbichler Co. Ltd.). Right: the modern PV technology is integrated into a traditional context (Ing. Studio Blasbichler Co. Ltd.)

**Energy integration**

The BIPV system is estimated to produce 41,327 kWh per year. It feeds the total generated electricity into the grid together with second PV plant integrated on a nearby building, of the same owner. A solar thermal installation is integrated close to the second PV plant (Ing. Studio Blasbichler Co. Ltd.).

**Technology integration**

188 Scheuten Multisol Vitro (P6-54) photovoltaic modules are integrated on the southeast facing roof of the building. The modules are frameless glass–glass products. The polycrystalline cells are inserted between a highly transparent low-iron-tempered safety glass, with anti-reflective-coating (front), and a heat strengthened safety glass (rear). The cells do not cover the whole module area leaving gaps to let light through (Fig. 3.24, left). The glazed PV modules are mounted with special aluminium Solrif profile frames and fixed to the wood substructure (Fig. 3.24, right). Special mounting clamps brace two modules to the frames in the overlapping area, ensuring the system's weather tightness (Fig. 3.25, left).

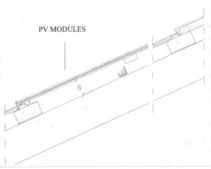

**Fig. 3.24** Left: wood structure supporting the BIPV plant (Ing. Studio Blasbichler Co. Ltd.). Right: technical detail of the modules fixing system, re-drawn by Eurac research (Ing. Studio Blasbichler Co. Ltd.)

**Fig. 3.25** Left: detailed view of the 'Solrif' mounting system (Ing. Studio Blasbichler Co. Ltd.). Right: BIPV shading effect (Ing. Studio Blasbichler Co. Ltd.)

## DECISION-MAKING

When the owner decided to integrate a photovoltaic system into the farm building, he wanted to exploit the building structures' available surface in order to produce renewable energy to be fed into the grid and indirectly guarantee coverage of the building's energy consumption. He was also encouraged by the possibility of receiving economic incentives (Conto Energia) (Ing. Studio Blasbichler Co. Ltd.).

## PROCESS

The owner Thomas Widmann was supported from the first design stages by the engineering firm Blasbichler in finding the best architectural solution to apply to the existing buildings. The Blasbichler team provided a preliminary economic assessment and was the responsible for the BIPV technical design. Scheuten Solar BV supplied the modules. Elektro Lahner GmbH and Solarxpert Srl were involved in the plant installation, completed in 2011.

## LESSON LEARNT

The BIPV modules create a semi-transparent surface able to partially shade the upstairs barn (Fig. 3.25, right). They allow a fair amount of sunlight to penetrate, guaranteeing natural illumination and contributing to the building heating. However, they prevent an excessive solar gain. In this case study, the potential of the BIPV multifunctional technology is highly exploited. All the functions of a traditional roofing system (e.g. mechanical resistance, thermal insulation, protection from atmospheric agents, water tightness, etc.) are connected with the shading function that controls the internal visual and thermal comfort, without compromising the electric energy generation. It is an interesting example of BIPV application on an old construction, which is located in a high-value natural environment.

| BIPV SYSTEM DATA | |
| --- | --- |
| PV MODULES | Standard |
| SOLAR TECHNOLOGY | Polycrystalline silicon |
| NOMINAL POWER | 39.6 kWp |
| SYSTEM SIZE | 346 m$^2$ |
| MODULE SIZE | 1,488 × 988 mm |
| ORIENTATION | South-east |
| TILT | 35° |

| BIPV SYSTEM COSTS | |
| --- | --- |
| TOTAL COST (€) | 146,202 |
| €/m$^2$ | 423 |
| €/kWp | 3,701 |

| PRODUCER DATA | |
| --- | --- |
| PRODUCER | Scheuten Solar BV |
| ADDRESS | Hulsterweg 82, Venlo (NL) |
| CONTACT | info@scheutensolar.com |
| WEB | www.scheutensolar.com |

## 3.2.6   Enzian Office, Bolzano (BZ)

| PROJECT DATA | |
| --- | --- |
| PROJECT TYPE | New construction |
| BUILDING FUNCTION | Office |
| INTEGRATION SYSTEM | Opaque cold façade, semi-transparent façade |
| LOCATION | Via Ressel 3, Bolzano (BZ) |

### BUILDING/SYSTEM INTEGRATION

#### Aesthetic integration

Enzian Office is a 10-storeys building located in the industrial zone of Bolzano. The whole building is covered with photovoltaic modules integrated into the building glass facade (Fig. 3.26). The PV modules are made of amorphous silicon that homogenizes the external surfaces, so that the difference between opaque and semi-transparent facade parts is not recognizable (Fig. 3.28, left). The integrated PV skin makes the 'sustainable design' highly visible from outside (Fig. 3.27, left).

**Fig. 3.26** Enzian Office BIPV system: the modules replace opaque parts of the façade (under the windows hight) and semi-transparent sections (beside the windows) (Eurac research)

**Fig. 3.27** Left: the building railings also are made of glass BIPV modules (Eurac research). Right: detailed view of the semi-transparent modules texture (Eurac research)

### Energy integration

The building is certified CasaClima Gold. The PV system integrated into the building envelope, together with modules placed on the roof, produce around 113 MWh/year, supplying enough energy to feed the building's heating and cooling needs using a reversible heat pump and a pellet heating system. The system is grid connected, so the excess energy is fed into the power grid (Eurac research).

## Technology integration

According to the solar exposure of the building facades, either double or triple insulating glass with amorphous silicon modules or opaque laminated glass is used. The photovoltaic modules (Voltarlux) are designed on the basis of Schott Solar's ASI THRU thin-film technology as silicon tandem cells (3 mm) on a glass substrate (Fig. 3.27, right). Some modules replace the semi-transparent facade part. The interior is protected with laminated safety glass. The chamber between the glass panes is filled with argon for thermal insulation. Other modules replace the opaque facade part with an insulating layer behind them (Fig. 3.28, right). The gap between the modules and the insulation is 5 cm and is covered on the bottom and top. Cables are contained within the framing system.

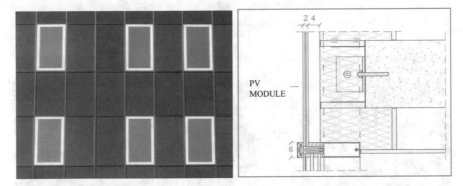

**Fig. 3.28** Left: external view of the modules metal framing system (Leitner Electro Srl). Right: technical detail of the modules fixing structure (opaque facade section), re-drawn by Eurac research (Leitner Electro Srl)

## DECISION-MAKING

The building was designed to be an energy self-sufficient unit. Hence, the decision to integrate a photovoltaic plant. The wide building facades were covered as much as possible with PV modules in order to maximize the electric energy production exploiting most of the available solar radiation. Amorphous silicon was chosen instead of crystalline silicon, because of its uniform shading effect inwards and its uniform appearance outwards (Energytech Srl). Additional PV modules were applied to the building roof in order to increase the electric energy building supply.

## PROCESS

The building construction was commissioned by the PSZ di Gral Srl. The design process involved several stakeholders, as Arch. Zeno Bampi and Ind. Eng. Franz Steiner, building designers, Eng. Sigfried Pohl, responsible for the structural assessment and works supervisor, Energytech Srl for the thermal plants, Leitner Electro Srl for the electric plants and the BIPV system design (being supported by the online PVGIS software). Glaswerke Arnold GmbH was chosen as PV modules supplier, as a result of a market research. The BIPV installation involved Leitner Electro Srl and Kaser GmbH, and was completed in 2011. Pohl Immobilien is the building owner.

## LESSON LEARNT

The PV modules are integrated into different building components, providing examples of how the PV might be used in place of traditional building materials. The PV substitutes the semi-transparent parts, the insulated windows, the external parapets and the external cladding (Fig. 3.29, right). In the semi-transparent part, it is used as a sun shielding without the need for additional shading provisions that would have increased the costs. Moreover, the amorphous silicon texture produces a special lighting scenario, a uniform shading effect that does not disturb the office's users (Fig. 3.29, left). The light controlling function of the photovoltaic cells is added to the insulating function of the glazing system, highlighting the multifunctional feature of the BIPV technology. Regarding the BIPV system design, one of the main challenges reported by the designer is related to the strict fire-safety regulations which need to be respected in the facade design (Energytech Srl).

**Fig. 3.29** Left: sun shielding effect of the BIPV modules (Eurac research). Right: the building shows an impressive appearance highly visible from the surrounding areas (Eurac research)

| BIPV SYSTEM DATA | |
|---|---|
| **PV MODULES** | Custom made |
| **SOLAR TECHNOLOGY** | Thin-film amorphous silicon |
| **NOMINAL POWER** | 100 kWp |
| **SYSTEM SIZE** | 2,340 m$^2$ |
| **MODULE SIZE** | 1,020 × 626 mm |
| **ORIENTATION** | West, South, East |
| **TILT** | 90° |

| BIPV SYSTEM COSTS | |
|---|---|
| **TOTAL COST (€)** | / |
| **€/m$^2$** | / |
| **€/kWp** | / |

| PRODUCER DATA | |
| --- | --- |
| **PRODUCER** | Arnold Glas GmbH |
| **ADDRESS** | Alfred Klingele Str. 15, Remshalden (D) |
| **CONTACT** | / |
| **WEB** | / |

## 3.2.7 La Pedevilla Chalet, Marebbe (BZ)

| PROJECT DATA | |
| --- | --- |
| **PROJECT TYPE** | New construction |
| **BUILDING FUNCTION** | Residential |
| **INTEGRATION SYSTEM** | Opaque tilted roof |
| **LOCATION** | Strada Pliscia 13, Marebbe (BZ) |

### BUILDING/SYSTEM INTEGRATION

#### Aesthetic integration

The photovoltaic system is integrated into the wooden roof of a mountain chalet (Fig. 3.30). It is made of black modules replacing the conventional roof boards, in a

**Fig. 3.30** La Pedevilla Chalet BIPV roofing system (Arch. Armin Pedevilla)

way that the plant blends in very well with the dark roof (Fig. 3.31, right). The building is surrounded by the impressive scenery of the South Tirol dolomite mountain ridges (Fig. 3.31, left).

**Fig. 3.31** Left: building embedded in the natural surroundings of Pliscia (Arch. Armin Pedevilla). Right: dark BIPV modules (Leitner Electro Srl)

### Energy integration

The chalet is certified 'CasaClima A' and achieved the nZEB (nearly Zero-Energy Building) target, due to the exploitation of solar and geothermal energy and the employment of energy-efficient building solutions. The photovoltaic system was calculated to produce around 6,592 kWh per year. It is able to cover the building's electric consumption for the ventilation system and the heat recovery. Around an 80% of the produced photovoltaic electricity is self-consumed. The system has access to the net metering scheme (Scambio sul Posto) (Leitner Electro Srl).

### Technology integration

The 25 photovoltaic modules (Aleo Solar S_79 SOL) are mounted with special aluminium Solrif profile frames and fixed to the substructure with special mounting clamps, used to brace two modules to their frames in the overlapping area (Fig. 3.32,

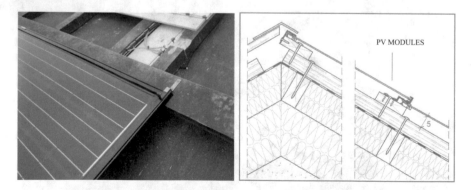

**Fig. 3.32** Left: detailed view of the 'Solrif' mounting system (Leitner Electro Srl), Right: technical detail of the roof bearing structure, re-drawn by Eurac research (Leitner Electro Srl)

left). This system ensures weather tightness thanks to the special horizontal interlock of the modules' profile frames and the additional rubber lip placed between the overlapping module edges. The modules are naturally back ventilated, due to the distance between the modules and the roof (Fig. 3.32, right). For safety reasons, the wiring and the connectors are placed in a grid gutter, fixed on fire-resistant plasterboard that separates the modules from the wooden parts of the roof (Fig. 3.33, left).

**Fig. 3.33** Left: wood modules substructure: the cables and the fire-resistant plasterboard are visible (Leitner Electro Srl). Right: the buildings evoke the old 'Paarhof' structure (Arch. Armin Pedevilla)

### DECISION-MAKING

From the beginning, the client included the PV plant into the building project, driven by the purpose of creating a high-energy efficiency building, able to produce the energy required using most of the on-site available RES (renewable energy sources). Once the economic feasibility was verified through simplified calculations, he confirmed the installation of the plant. The PV panels were architecturally integrated into the roof with the main goal of reaching a high level of aesthetic quality, matching the building's dark painted oak exterior (Leitner Electro Srl) (Fig. 3.33, right).

### PROCESS

The client, Arch. Armin Pedevilla, was personally involved in project process, being the main designer of the building. He was supported by the Bergmeister engineering firm for the structural studies. Leitner Electro Srl provided an offer for a photovoltaic plant and evaluated a first price calculation through a payback simulation and was chosen to be responsible for the PV system's architectural and technical design as well as the installation. Aleo Solar GmbH was chosen as the module supplier. In 2013, the building construction was concluded and the PV plant was connected to the electric grid.

## LESSON LEARNT

The construction can be considered a modern interpretation of the old 'Paarhof', the traditional kind of farm building typical for South Tyrol, built with a local natural material such as pine wood and white Dolomite concrete (Leitner Electro Srl). The photovoltaic technology is integrated combining contemporary with traditional design. Simple standard PV modules are placed on the roof as if they were conventional roofing components blended into the dark wood boards, making the building a great example of photovoltaic integration with high replication potential. This example shows that reaching an overall high quality of a BIPV system requires a careful design, with care for details and sensitivity to the surrounding, but it does not always require specific custom-made products. Since 'standard' products were used, the cost of this BIPV system (i.e. 2,603 €/kWp) is not much higher compared to 'ground mounted' systems.

| BIPV SYSTEM DATA | |
| --- | --- |
| PV MODULES | Standard |
| SOLAR TECHNOLOGY | Monocrystalline silicon |
| NOMINAL POWER | 6 kWp |
| SYSTEM SIZE | 43.3 m$^2$ |
| MODULE SIZE | 1,016 × 1,704 mm |
| ORIENTATION | South-east |
| TILT | 30° |

| BIPV SYSTEM COSTS | |
| --- | --- |
| TOTAL COST (€) | 15,616 |
| €/m$^2$ | 361 |
| €/kWp | 2,603 |

| PRODUCER DATA | |
| --- | --- |
| PRODUCER | Aleo Solar GmbH |
| ADDRESS | Marius Eriksen Strasse 1, Prenzlau (D) |
| CONTACT | info@aleo-solar.de |
| WEB | www.aleo-solar.com |

## 3.2.8 Hafner Energy Tower, Bolzano (BZ)

| PROJECT DATA | |
| --- | --- |
| PROJECT TYPE | New construction |
| BUILDING FUNCTION | Office |
| INTEGRATION SYSTEM | External opaque device (shading) |
| LOCATION | Via Negrelli 5, Bolzano (BZ) |

## BUILDING/SYSTEM INTEGRATION

### Aesthetic integration

Hafner Energy Tower is a 13-storeys triangular building standing close to the South Bolzano highway exit (Fig. 3.34). On the south side, a slanted PV-sail is installed over the whole height of the building, providing a permanent shading structure (Fig. 3.35, left). On the top, it seems to intersect the prism structure, creating an aesthetically singular composition, where every element is functionally designed towards energy efficiency (Fig. 3.35, right).

**Fig. 3.34**  Hafner Tower BIPV sail (Eurac research)

**Fig. 3.35**  Left: the photovoltaic sail shades the South front of the tower (Eurac research). Right: the sail structure intersects the building block (Arch. Fabio Rossa)

**Energy integration**

The PV system integrated into the sail was estimated to produce around 45,690 kWh (Eurac research) yearly, with a self-consumption rate of 60% (Arch. Fabio Rossa). The PV production supplies the building's heating and cooling systems using a reversible heat pump. It is able to provide the 26% of the total electric load, with the remaining load covered by a biomass-driven cogeneration plant, a small wind turbine and a rotor that exploits the air flow ascending through the double skin facades (Eurac research). These technologies broadly contribute the building's to the 'Zero Energy' and CasaClima A certifications.

**Technology integration**

The BIPV plant is made of standard dark opaque modules (EGM series) distributed on the whole sail. The modules are rack mounted, fixed through horizontal tubular profiles to a metal grate reinforced by reticular elements (Figs. 3.36, left and right). The main sail structure is strengthened at the base and anchored to the building block through metal beams (Fig. 3.37, left).

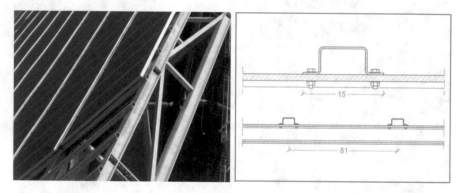

**Fig. 3.36** Left: construction phase: the BIPV mounting structure is still visible (Eurac research). Right: technical detail of the modules fixing system, re-drawn by Eurac research (Arch. Fabio Rossa)

**Fig. 3.37** Left: metal beams anchoring the sail to the building (Arch. Fabio Rossa). Right: fake modules aesthetically match the BIPV plant surface (Eurac research)

## DECISION-MAKING

Supporting the energy sustainability by the photovoltaic technology exploitation was one of the main purposes of Hafner Service Srl energy policy. The Hafner Tower was conceived from the beginning as a $CO_2$-neutral building, able to exploit available renewable energy sources (Arch. Fabio Rossa). The building plan, including the photovoltaic sail which was outlined in the first design phases, was highly appreciated by the public administration.

## PROCESS

In 2007, Hafner Service Srl commissioned the new tower design to Arch. Fabio Rossa and Stefano Dalprà, of the Area 17 firm. The building design and construction process involved several stakeholders, as Eng. Martin Weiss (H&T Planungsbüro), also coordinator and works supervisor, Eng. Wolfram Sparber (Eurac research), responsible for the energy assessment, Graf AG, PV system installer. The Chinese Changzhou EGing PV Tech. Co. Ltd. supplied the BIPV modules. The main construction works were concluded in 2011.

## LESSON LEARNT

In the Hafner Energy Tower, the photovoltaic system is integrated as a multifunctional structure providing not only electric energy but also a shading function. The PV-sail protects large building glazed surfaces from an excessive solar gain, decreasing the cooling energy demand. It is integrated slanted on the building block. The intersection point, where the tower and the sail cross, needed a complex design. A few fake modules were installed in order to fit the gap perfectly (Fig. 3.37, left). They complete the homogeneous PV surface that covers the sail, conferring to the whole structure an aesthetically interesting result.

| BIPV SYSTEM DATA | |
|---|---|
| PV MODULES | Standard |
| SOLAR TECHNOLOGY | Monocrystalline silicon |
| NOMINAL POWER | 90.46 kWp |
| SYSTEM SIZE | 601.3 m$^2$ |
| MODULE SIZE | 1,580 × 808 mm |
| ORIENTATION | South |
| TILT | 88° |

| BIPV SYSTEM COSTS | |
|---|---|
| TOTAL COST (€) | 320,000 |
| €/m$^2$ | 532 |
| €/kWp | 3,538 |

| PRODUCER DATA | |
| --- | --- |
| PRODUCER | Changzhou EGing PV Tech. Co., Ltd. |
| ADDRESS | Jinwu Road 18, Jintan (China) |
| CONTACT | / |
| WEB | www.egingpv.com |

## 3.2.9  Multi-family House, Appiano (BZ)

| PROJECT DATA | |
| --- | --- |
| PROJECT TYPE | Retrofit |
| BUILDING FUNCTION | Residential |
| INTEGRATION SYSTEM | Opaque tilted roof |
| LOCATION | Appiano (BZ) |

## BUILDING/SYSTEM INTEGRATION

### Aesthetic integration
The PV system is integrated on four dormers, on the western roof slope of a residential building built in 1977 in the historical town centre of Appiano (Fig. 3.38). It represents an interesting retrofit installation where the dark surfaces of the photovoltaic modules are installed next to the traditional roof tiles (Figs. 3.39, left and right). This contrast is not visible from the surrounding building or from the street (building owner).

Fig. 3.38  Multi-family house BIPV roofing system (building owner)

**Fig. 3.39** Left: aerial view of the building: the modules contrast with the traditional surrounding roofing systems (building owner). Right: the building roof is higher than the other ones so that there is no evidence of the modules from the surrounding (building owner)

**Energy integration**

With a nominal power of 1.92 kWp, the BIPV system covers almost the 70% of the top apartment's electricity demand thanks to an annual energy production of around 1,900 kWh. The photovoltaic output is especially used for the building conditioning system (building owner).

**Technology integration**

The BIPV plant is composed of 32 thin-film amorphous silicon standard modules (Kaneka K60). The modules are installed on aluminium planks which are fixed on the dormers' metal sheets through special grab connectors (Figs. 3.40, left and right). This mounting system does not require drilling the metal sheets, ensuring the roofs weather tightness. The natural ventilation of the PV plant is guaranteed thanks to an air gap between the modules and the roof (Fig. 3.41, left).

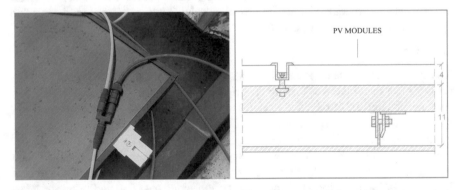

**Fig. 3.40** Left: detailed view of the modules cables and the special fixing grab (building owner). Right: technical detail of the BIPV mounting system, re-drawn by Eurac research (Phys. Francesco Nesi)

**Fig. 3.41** Left: the modules are distanced from the dormers metal sheets (PVEnergy). Right: the BIPV plant is slightly tilted: this requires high care for snowfalls (building owner)

## DECISION-MAKING

The building owner decided to install a photovoltaic plant in order to balance the increase of the electricity demand following the installation of a new air-conditioning system. He designed the PV plant as a distinctive in-roofing system, using the existing dormers as support. The purpose was to hide the modules from view, covering the dormers with the same shape and size. In this way, the difference between the plant surfaces and the roof tiles is not noticeable (building owner).

## PROCESS

The building's owner was involved during all process phases, as responsible for the architectural and technical design of the BIPV system. After performing an overall economic assessment and a market survey, he selected the modules. PVEnergy Srl, already known to the client thanks to previous collaborations, was chosen to install the PV plant. Kaneka Corporation is the manufacturer of the modules. The installation was completed in 2007.

## LESSON LEARNT

The PV integration had good results concerning several aspects. Aesthetically, the designer met the objectives to keep the original homogeneous surface of the pitched roof by placing uniformly the modules on the dormers. A better aesthetic integration could have been performed by closing the air gaps, but the modules are hidden from view so it was not necessary to do so (Fig. 3.41, right). Economically, the selected module technology (with low efficiency) allowed the photovoltaic plant to stay within the small plant's category of the Italian energy service management body (GSE). Technically, after the construction work was finished, a minor intervention was required in order to avoid snow damage on the roof gutter (building owner). As a multifunctional technology, BIPV should be designed taking several targets into account (e.g. the ones mentioned above) which have to be balanced according to the main project purposes.

| BIPV SYSTEM DATA | |
|---|---|
| PV MODULES | Standard |
| SOLAR TECHNOLOGY | Thin-film amorphous silicon |
| NOMINAL POWER | 1.92 kWp |
| SYSTEM SIZE | 30 m$^2$ |
| MODULE SIZE | 960 × 990 mm |
| ORIENTATION | West |
| TILT | 8° |

| BIPV SYSTEM COSTS | |
|---|---|
| TOTAL COST (€) | 10,000 |
| €/m$^2$ | 333 |
| €/kWp | 5,208 |

| PRODUCER DATA | |
|---|---|
| PRODUCER | Kaneka Corporation |
| ADDRESS | Nakanoshima 2-3-18, Osaka (JP) |
| CONTACT | / |
| WEB | www.kaneka-solar.com |

## 3.2.10  Autobrennero Headquarter, Trento (TN)

| PROJECT DATA | |
|---|---|
| PROJECT TYPE | Retrofit |
| BUILDING FUNCTION | Office |
| INTEGRATION SYSTEM | Opaque cold façade |
| LOCATION | Via Berlino 10, Trento (TN) |

## BUILDING/SYSTEM INTEGRATION

### Aesthetic integration

The headquarters of the company Autobrennero is located at the motorway exit
Trento Centro. The building underwent substantial refurbishment, where the
existing 5-floor building was extended and characterized by a curtain wall system in
the new façade. Dark photovoltaic modules were integrated as building stringcourses on one of the three sides of the building which are covered with a double
skin façade system (Fig. 3.42). They take up the features of the pre-existing block
aesthetically and unify it with the new one (Figs. 3.43, left and right).

**Fig. 3.42** Autobrennero headquarter BIPV system as dark building stringcourses (FAR Systems SpA)

**Fig. 3.43** Left: the new building portion with the facade BIPV system is visible on the left side of the existing block (white roof) (FAR Systems SpA). Right: the modules appearance matches the existing building dark colours (FAR Systems SpA)

**Energy integration**

The BIPV energy production is used to power the building's electrical services, especially the ventilation system. It was measured to cover about a 4% of the electricity demand (Autobrennero SpA). A geothermal system supports the building's thermal energy need. The renewable energy production is introduced into a structure already designed to save the energy. The double skin façade containing the PV modules is a high performance, advanced technology, covering an area of 720 m$^2$ and interacts with the heating and cooling plant.

**Technology integration**

The BIPV plant is made of 128 black/transparent CIS (Copper Indium Selenide) modules with 20% transparency (Fig. 3.44, left). The modules are integrated between high insulation double glazing panels. Behind the transparent PV, an insulating layer is placed in order to prevent an excessive increase in temperature (Fig. 3.44, right). The modules are secured on three sides by aluminium profiles with thermal cutting, and silicone sealant along the fourth edge, where they touch other modules (Fig. 3.45, left). Due to the glazed PV module's semi-transparency, the junction box is visible from outside.

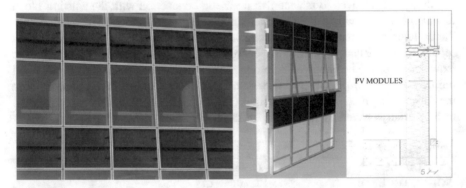

**Fig. 3.44** Left: detailed view of the BIPV facade: the semi-transparent modules make the junction boxes visible from outside (FAR Systems SpA). Right: technical detail of the BIPV mounting system (opaque facade section), re-drawn by Eurac research (FAR Systems SpA)

**Fig. 3.45** Left: fake modules installed, matching the modules texture (FAR Systems SpA). Right: the building is placed in a mountain landscapes that influences the BIPV output (FAR Systems SpA)

**DECISION-MAKING**

The integration of the PV technology into the building envelope arose from a technical opportunity. The Autobrennero SpA General Technical Division decided to exploit the particular structure of the double skin facade planned into the building's refurbishment project (Autobrennero SpA). This intervention is mainly related to the growing commitment towards a sustainable environment.

**PROCESS**

Autobrennero SpA, the owner society, commissioned the building refurbishment design to Arch. Sergio Giovanazzi. The BIPV system realization involved Tosoni Group SpA, a specialist in curtain walls for buildings, and Far Systems SpA, responsible for the architectural and technical design of the BIPV plant, as well as modules installer. The modules, provided by Würth Solar GmbH, were mounted in 14 days only. The works were completed in 2009.

**LESSON LEARNT**

From an aesthetic point of view, choosing the thin-film CIS photovoltaic technology created homogeneous surfaces. The modules' texture resulted to be similar to the glass. It allowed to maintain an aesthetic connection with the existing block and also gave an architectural improvement to the building fronts. The relation with the building's surrounding was essential. A BIPV system should be designed with regard to its context as well as its energy performance, in order to achieve the expected results. The photovoltaic production was first estimated through the r.sun numerical model of PVGIS webGIS system. It was evaluated taking into account the surrounding mountainous landscapes (Fig. 3.45, right). The installation resulted to be partially shaded by mountains on the horizon (FAR Systems SpA).

| BIPV SYSTEM DATA | |
| --- | --- |
| PV MODULES | Custom made |
| SOLAR TECHNOLOGY | Thin-film CIS |
| NOMINAL POWER | 7 kWp |
| SYSTEM SIZE | 97 m$^2$ |
| MODULE SIZE | 1,170 × 600 mm |
| ORIENTATION | East |
| TILT | 90° |

| BIPV SYSTEM COSTS | |
| --- | --- |
| TOTAL COST (€) | / |
| €/m$^2$ | / |
| €/kWp | / |

| PRODUCER DATA | |
| --- | --- |
| PRODUCER | Würth Solar GmbH |
| ADDRESS | Alfred Leikam Str. 25, Schwäbisch Hall (D) |
| CONTACT | wuerth-solar@we-online.de |
| WEB | www.wuerth-solar.de |

## 3.2.11   Ex-Post, Bolzano (BZ)

| PROJECT DATA | |
| --- | --- |
| PROJECT TYPE | Retrofit |
| BUILDING FUNCTION | Office |
| INTEGRATION SYSTEM | Opaque cold façade |
| LOCATION | Via Renon 3, Bolzano (BZ) |

### BUILDING/SYSTEM INTEGRATION

#### Aesthetic integration

The BIPV plant covers the south-east facing staircase of the 5-storey former post building (Fig. 3.46). Built in 1954, it was substantially renovated for the relocation

**Fig. 3.46** Ex-Post BIPV system (Eurac research)

of the Environmental Department offices. Dark modules create a large opaque surface, interrupted on the most visible façade with a stripe of semi-transparent glass panels to allow natural illumination of the staircase behind (Figs. 3.47, left and right). The integrated PV system seems to create a tower next to the other white facades and provide a singular aspect to a visible building, placed in the city centre of Bolzano.

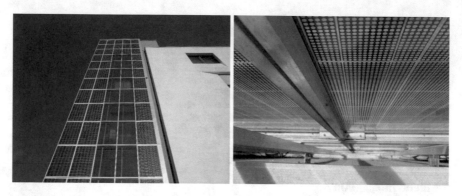

**Fig. 3.47** Left: a stripe of semi-transparent glass panels allow the natural illumination of the staircase behind (Eurac research). Right: view from inside the staircase: the mounting system of the glass panels is visible (Eurac research)

**Energy integration**

The building was calculated with PHPP (passive house planning programme, provided by Passivhaus Institut Darmstadt) to an energy demand of 10 kWh/m$^2$a. Furthermore, it was certified as CasaClima Gold according to the regional rating scheme, with a calculated energy demand of 7 kWh/m$^2$a (Eurac research). The BIPV system is one of the main strategies implemented in order to decrease the energy demand, as efficient plants and envelope components. According to the system monitoring results (Eurac research), the PV production contributes its part to the demand, but can rarely make up for it completely. Considering a monthly electric energy balance in a period between June 2011 and April 2012, it cannot meet the building's electric energy demand in any season, resulting to be between 4.5 and 12 times lower than the building electric load at the same time. Considering hourly values, for 5% of the time an electric energy overproduction occurs. It could theoretically be re-used through the optimization of a self-consumption system.

**Technology integration**

The Ex-Post BIPV plant is made of 162 Schott Solar PV modules (ASE-165-GT-FT/MC), applied on the existing façade through a metallic structure as a cladding of the wall (Fig. 3.48, left). The modules are not retro-ventilated because, despite a 15 cm gap between the modules and the wall (Fig. 3.48, right). A frame surrounding the PV system obstructs the air flow in the gap (Fig. 3.49, left).

**Fig. 3.48** Left: steel system fixing the BIPV plant to the existent building (Elpo GmbH). Right: the modules are distanced from the concrete building façade (Elpo GmbH)

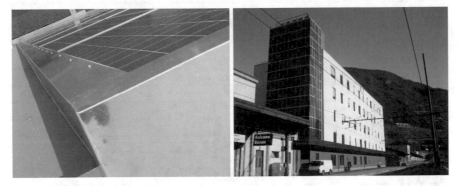

**Fig. 3.49** Left: detailed view of the BIPV plant surrounding frame (Elpo GmbH). Right: view from Bolzano railway station (Eurac research)

## DECISION-MAKING

The retrofit designer introduced the idea to integrate PV modules. As the main purpose of the renovation was energy efficiency he decided to include PV technology in the building. The plant was integrated on two highly visible facades which face the railway, as a symbolic evidence of the high commitment spent by the Autonomous Province of Bolzano towards a sustainable direction (Fig. 3.49, right). Integrating architectural ambition, energy efficiency and innovation, the BIPV system can increase the building's value (Arch. Michael Tribus).

## PROCESS

The Autonomous Province of Bolzano commissioned Arch. Michael Tribus with the building retrofit design. The refurbishment focused on the envelope redefinition and the building plants renovation, in order to reduce the heating demand. In 2005, once the works were started, the integration of the PV system was approved by the client. Schott Solar AG was chosen as modules supplier. Obrist Gmbh and Elpo Gmbh were involved with the electric plant design and the BIPV system installation.

**LESSON LEARNT**

Thanks to the BIPV system monitoring operations (Eurac research), some aspects, which represent typical problems for BIPV located in urban environments were discovered. The operation of the system resulted to be compromised from shading causing an energy loss of around 13%. Furthermore, the photovoltaic output was evaluated to decrease another 4% because of the lack of back ventilation of the modules, which reach temperatures of up to 57.4 °C in summer (Eurac research). This experience highlights that if dealing with BIPV, the surrounding context is one of the main components to evaluate. The landscape and urban morphology have a substantial shading effect that can highly compromise the photovoltaic energy performance, as much as the weather conditions.

| BIPV SYSTEM DATA | |
| --- | --- |
| PV MODULES | Standard |
| SOLAR TECHNOLOGY | Polycrystalline silicon |
| NOMINAL POWER | 26.7 kWp |
| SYSTEM SIZE | 212 m$^2$ |
| MODULE SIZE | 1,620 × 810 mm |
| ORIENTATION | South-west, South-east |
| TILT | 90° |

| BIPV SYSTEM COSTS | |
| --- | --- |
| TOTAL COST (€) | 222,000 |
| €/m$^2$ | 1,047 |
| €/kWp | 8,315 |

| PRODUCER DATA | |
| --- | --- |
| PRODUCER | Schott Solar AG |
| ADDRESS | Hattenbergstrasse 10, Mainz (D) |
| CONTACT | / |
| WEB | www.schott.com |

## 3.2.12   Le Albere District, Trento (TN)

| PROJECT DATA | |
| --- | --- |
| PROJECT TYPE | New construction |
| BUILDING FUNCTION | Residential, office |
| INTEGRATION SYSTEM | External semi-transparent device |
| LOCATION | Via San Severino, Trento (TN) |

## BUILDING/SYSTEM INTEGRATION

**Aesthetic integration**

The new district Le Albere, built from restructuring a former industrial area in Trento, is primarily characterized by its innovative urban fabric (Fig. 3.50). The district includes commercial, residential and office buildings. A large surface of photovoltaic modules is integrated into the buildings, representing one of the most important and unifying features of the entire project (Fig. 3.51, left). With different buildings heights and inclinations, the district is harmoniously inserted within its surrounding mountain environment (Fig. 3.51, right).

**Energy integration**

The residential and office structures are designed as 'passive' buildings and certified according to CasaClima standards. The building's energy provision is guaranteed by systems which exploit different renewable energy sources. The BIPV plant is divided into 11 sub-plants, which are independently connected to the electricity grid. The PV production supplies part of the electrical energy demand of offices, common spaces, pump rooms and the basement areas lighting (e.g. staircases and the district park). A trigeneration plant and an autonomous geothermal system provide energy for heating and cooling of the whole district buildings.

**Fig. 3.50** Le Albere BIPV external device hanging over the buildings' roof (Eurac research)

**Fig. 3.51** Left: the modules are placed on the most of the district buildings as a unifying element (FAR System Srl). Right: the buildings follow the surrounding mountain environment with different heights and inclinations (Eurac research)

## Technology integration

The BIPV plants are made of different typologies of custom-made modules with a silver appearance (Figs. 3.52, left and right). 4,160 glass/Tedlar modules are mounted on metal frames (a), which are made of two metal struts of rectangular section and three of hollow circular section. 985 glass/glass modules are anchored to the roofs metal sheets with special clamps and reinforced with a metal profile along the larger side (b). The mounting system consists of extremely lightweight extruded aluminium profiles. It is flexible thanks to the customized sliding mechanism that balances the component's thermal expansions (FAR Systems Srl).

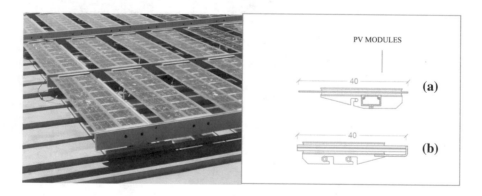

**Fig. 3.52** Left: glass/Tedlar modules (**a**) mounted on metal frames fixed to the buildings' roof (FAR System Srl). Right: technical details of the two different typologies of modules structure and mounting system: glass/Tedlar modules (**a**) and glass/glass modules (**b**) re-drawing of Eurac research (FAR System Srl)

## DECISION-MAKING

The project started with a partnership among public institutions, private companies and asset management companies. The project was conceived from the beginning as an educative instrument aimed towards the energy saving and an environmentally responsible management [2].

## PROCESS

Once the clients Castello SGR and Itas Assicurazioni bought the former Michelin area, they committed the general district design to Renzo Piano Building Workshop Srl (2002). Several stakeholders (designers, technicians, consultants, manufacturers) were involved within the district development. FAR Systems Srl was mainly responsible for the BIPV systems design and installation, working with Iure Srl for the project management. The works were concluded in 2013.

## LESSON LEARNT

The 'aesthetic' integration is one of the main issues in this case study. It shows how a well-known architect (Renzo Piano) decided to use the PV modules to shape the buildings aesthetic, clearly declaring their presence and making them highly visible instead than hiding or camouflaging. The architects used the PV modules to express their architectural language, as a key element in the whole building composition (Figs. 3.53, left and right).

**Fig. 3.53** Left: glass/glass modules (**b**) anchored with special clamps to existing structures (FAR System Srl). Right: the BIPV system is highly visible as a key element in the whole building composition (FAR System Srl)

Regarding the BIPV system design, one of the main challenges was to customize the PV modules in order to satisfy the aesthetic requirements of the architects in terms of colours, dimensions, semi-transparency and materials as well as keeping costs in an acceptable range (FAR Systems Srl).

| BIPV SYSTEM DATA | |
| --- | --- |
| PV MODULES | Custom made |
| SOLAR TECHNOLOGY | Polycrystalline silicon |
| NOMINAL POWER | 279 kWp |
| SYSTEM SIZE | 3,258 m$^2$ |
| MODULE SIZE | 1,600 × 400 mm, 1,045 × 400 mm |
| ORIENTATION | South, West, East |
| TILT | 5°, 7.5°, 15° (a)–12° (b) |

| BIPV SYSTEM COSTS | |
| --- | --- |
| TOTAL COST (€) | 1,800,000 |
| €/m$^2$ | 552 |
| €/kWp | 6,450 |

| PRODUCER DATA | |
| --- | --- |
| PRODUCER | Guandong Golden Glass Techn. Co., LTD. |
| ADDRESS | Shantou College Road, Guangdong (CN) |
| CONTACT | info@golden-glass.cn |
| WEB | www.golden-glass.com |

## 3.2.13   Maso Lampele, Varna (BZ)

| PROJECT DATA | |
| --- | --- |
| PROJECT TYPE | New construction |
| BUILDING FUNCTION | Residential |
| INTEGRATION SYSTEM | Opaque tilted roof |
| LOCATION | Sottopaese 17, Varna (BZ) |

### BUILDING/SYSTEM INTEGRATION

#### Aesthetic integration

Maso Lampele is a residential structure consisting of three blocks. The owner lives in the structure and also runs his carpentry here. It is located at the base of the Eisack Valley's western slope (Fig. 3.54). The buildings harmoniously combine modern aspect (such as an unconventional shape, large windows and the photo-voltaic system integrated on two roofs) with the natural landscape from which the buildings take their morphology, using local materials (Figs. 3.55, left and right). The slightly slanted roofs reflect the shape of the surrounding hills. The buildings are constructed using mainly traditional materials, as natural stone and wood.

**Fig. 3.54** Maso Lampele BIPV roofing system (Arch. Norbert Dalsass)

**Fig. 3.55** Left: the buildings are integrated into the landscape morphology (Arch. Norbert Dalsass). Right: air gap between the building roof and the BIPV modules (Werner Graber)

**Energy integration**

Maso Lampele is designed as 'passive' building. It produces more energy than it uses. An annual electric energy demand of 16,700 kWh (including the building ventilation system consumption) is covered by the photovoltaic production and a surplus of 2,000 kWh is yearly fed into the electrical grid. The building's thermal energy demand, instead, is supplied using a wood burner powered by the waste of the owner's carpentry (Andreas Brunner).

**Technology integration**

The BIPV system is divided into two parts, one with a peak power of 20 kWp supplying the carpentry's electric demand and the second one with 10 kWp for the residential loads. It consists of 138 photovoltaic modules Kyocera (KD215GH-2PU). The modules are fixed with metal hooks to aluminium profiles. The mounting system is anchored with special clamps to the metal roofing surfaces (Figs. 3.56, left and right).

**Fig. 3.56** Left: the mounting system is anchored with special clamps to the metal roofing (Werner Graber). Right: detailed view of the modules and the mounting system (Werner Graber)

## DECISION-MAKING

The buildings' owner decided to create an energy-autonomous house, an example of ecological construction and energy efficiency. He wanted to include photovoltaic technology into the building's design without neglecting the aesthetic quality. From the first design phase, the building's profiles were defined in order to harmoniously integrate the PV modules and optimize their performance. The owner and the building's designer worked together to create an architecturally captivating structure (Andreas Brunner).

## PROCESS

During the whole buildings plan development, the owner Andreas Brunner worked together with the designers and the other specialists. He had not only the client role, but also he took part in the construction works as a carpenter. Arch. Norbert Dalsass was commissioned with the general building's design. Elektro Graber GmbH Company implemented the electric system and was technically responsible for the BIPV plants design and installation. The buildings were completed in 2012.

## LESSON LEARNT

Achieving a high aesthetical quality was one of the main goals of the project. Much of the roof metal sheets were covered with the photovoltaic plants in order to create uniform surfaces. The remaining parts were completed by placing wood boards, which take up the theme of the façade's feature (Andreas Brunner) (Fig. 3.57, left). The modern materials of the photovoltaic plants are integrated into an assembly of traditional building materials, typically used for the local constructions. Maso Lampele could be defined as a modern reproduction of the traditional building type highly common in the Alto Adige region and characterized by the harmonious interplay of local wood and natural stone (Fig. 3.57, right).

**Fig. 3.57** Left: wood boards are placed to complete the BIPV plant, taking the façade feature (Werner Graber). Right: modern reproduction of traditional houses made with local wood and natural stone (Arch. Norbert Dalsass)

| BIPV SYSTEM DATA | |
| --- | --- |
| **PV MODULES** | Standard |
| **SOLAR TECHNOLOGY** | Polycrystalline silicon |
| **NOMINAL POWER** | 30 kWp |
| **SYSTEM SIZE** | 205 m$^2$ |
| **MODULE SIZE** | 1,500 × 990 mm |
| **ORIENTATION** | South |
| **TILT** | 6° |

| BIPV SYSTEM COSTS | |
| --- | --- |
| **TOTAL COST (€)** | 130,000 |
| **€/m$^2$** | 634 |
| **€/kWp** | 4,333 |

| PRODUCER DATA | |
| --- | --- |
| **PRODUCER** | Kyocera Corporation |
| **ADDRESS** | Kyoto Japan |
| **CONTACT** | / |
| **WEB** | www.kyocera.co |

### 3.2.14 District Heating Plant, Laces (BZ)

| PROJECT DATA | |
|---|---|
| PROJECT TYPE | New construction |
| BUILDING FUNCTION | Industrial |
| INTEGRATION SYSTEM | Opaque cold façade, semi-transparent façade |
| LOCATION | Via Nazionale 2a, Laces (BZ) |

## BUILDING/SYSTEM INTEGRATION

### Aesthetic integration

The district heating plant is placed below a wooden slope of the Ötztal Alps. The photovoltaic system is integrated on the west- and south-facing facades (Fig. 3.58). It creates a dark envelope, which is aesthetically embedded in the surrounding landscape (Fig. 3.59, left). The modules comply with the irregular shape of the building (Fig. 3.59, right). The plan was designed based on the biomass burners (Eng. Klaus Fleischmann) and the fronts follow the sloped roof, where another PV plant is applied.

**Fig. 3.58** District heating plant BIPV façade system (Günther Wallnöfer)

**Fig. 3.59** Left: the irregular building shape complies with the surrounding mountain landscape (Google maps). Right: the plant covers the opaque parts as a cold façade system generating a stack effect (Eng. Klaus Fleischmann)

**Energy integration**

Together with the roof PV system, the integrated modules are measured to cover the 20–30% of the electricity building demand, mainly required by the burners (Eng. Klaus Fleischmann).

**Technology integration**

The integrated PV modules Solarwatt (M140-36 GEG LK XL) are standard semi-transparent glazed panels made of 36 monocrystalline cells (Fig. 3.60, left). On the building's west façade, they are fixed through aluminium profiles to the opaque concrete surface with a distance of 8 cm (Fig. 3.60, right). The air gap generates a stack effect that guarantees the modules natural ventilation. On the building's south façade, the modules are used as an exposure opening for the burner house, creating a warm façade directly exposed to the internal space. Thanks to a 10–20% transparency, they allow a natural illumination inside (Fig. 3.61, left). The same mounting system is fixed to a wooden structure.

**Fig. 3.60** Left: detailed view of the semi-transparent frameless modules (Günther Wallnöfer). Right: steel system supporting the modules that cover the opaque building facades (Günther Wallnöfer)

**Fig. 3.61** Left: view from inside the building: the supporting wood structure and the modules semi-transparency are visible (Günther Wallnöfer). Right: building West facade: the faked modules on the top follow the roof profile (Eng. Klaus Fleischmann)

## DECISION-MAKING

The building owner (EGL cooperative energy company of Laces) decided to equip the building with a PV plant to partially cover the electric demand of the biomass burners. This was inspired by the provincial programmes aimed to promote solar energy production (Eng. Klaus Fleischmann). The use of renewable energy, which is locally generated, decreases the costs and the environmental impact of the district heating plant. The initial idea was to build a plant on the roof. Later it was decided to install an additional plant on the façades, using the semi-transparent glass modules (Eng. Klaus Fleischmann).

## PROCESS

Public institutions (European, national and regional) played a crucial role in planning and partially financing the works. The mayor Karl Weiss, founder and chairman of EGL company, was strongly committed to the project. From the first stage of the building planning process, the Fleischmann & Jansen engineering firm was involved. It provided the structural design, an energy evaluation through ACCA software and an economic assessment of the PV system. Arch. Werner Pircher dealt with the architectural outline. Solarwatt Srl provided the photovoltaic modules. The PV system installation was commissioned to Wallnöfer Günther & Rudolf Snc and completed in 2009.

## LESSON LEARNT

On the south-facing facade, the photovoltaic modules are integrated also replacing transparent glazed surfaces. Other than performing the common functions of building envelope components, they are used as shading devices, guaranteeing the internal visual comfort. The same modules are integrated covering opaque surfaces to give a uniform appearance to the building facades. For the same reason, some fake modules were used (non-rectangular modules) on the top of the West facade, following the sloped roof profile (Fig. 3.61, right). During the construction works, some panels turned out not to have the correct size and were replaced with other ones. (Eng. Klaus Fleischmann)

| BIPV SYSTEM DATA | |
|---|---|
| PV MODULES | Standard |
| SOLAR TECHNOLOGY | Monocrystalline silicon |
| NOMINAL POWER | 48.9 kWp |
| SYSTEM SIZE | 446.7 m² |
| MODULE SIZE | 1,600 × 800 mm |
| ORIENTATION | South, West |
| TILT | 90° |

| BIPV SYSTEM COSTS | |
|---|---|
| TOTAL COST (€) | 357,416 |
| €/m² | 800 |
| €/kWp | 7,309 |

| PRODUCER DATA | |
|---|---|
| PRODUCER | Solarwatt Srl |
| ADDRESS | Maria Reiche Straße 2A, Dresda (D) |
| CONTACT | info@solarwatt.net |
| WEB | www.solarwatt.net |

## 3.2.15   Smart Lab, Rovereto (TN)

| PRODUCER DATA | |
|---|---|
| PROJECT TYPE | New construction |
| BUILDING FUNCTION | Public |
| INTEGRATION SYSTEM | Opaque cold façade |
| LOCATION | Viale Trento 46, Rovereto (TN) |

### BUILDING/SYSTEM INTEGRATION

#### Aesthetic integration

The BIPV plant consists of thin-film modules that cover almost a whole facade of Smart Lab, the youth centre of Rovereto (Fig. 3.62). The modules comprise a homogeneous surface. They have a significant visual impact (Fig. 3.63, left), increasing the architectural value of a building placed at the city gates. The building is designed for high visitor numbers. The building is a sociocultural meeting place for young people run by the local youth association.

**Fig. 3.62** Smart Lab BIPV façade system (Arch. Gianluca Perottoni)

**Fig. 3.63** Left: the BIPV facade is visibly exposed to the community (Arch. Gianluca Perottoni). Right: construction phase: the South building front behind the BIPV plant is still visible (Schüco International Italia Srl)

### Energy integration

The photovoltaic façade is a semi-transparent layer within a double skin system, able to control the solar gain (Arch. Gianluca Perottoni). It was calculated to produce around 9,072 kWh per year (PVGIS photovoltaic software), providing more than the 30% of the estimated building electricity demand (Schüco International Italia Srl). The BIPV system is one of the measures adopted in Smart Lab with the prospect of green building and energy efficiency (e.g. high quality of the technical-constructive strategies, recycled materials, etc.).

**Technology integration**

90 photovoltaic modules ProSol TF+ are integrated according to Schüco ventilated façade system. The modules are made of microamorphous silicon cells (20% transparency), which combine amorphous and microcrystalline silicon. The bearing structure is composed of metal uprights and crosspieces hiding the junction boxes and the cabling system (Figs. 3.64, left and right).

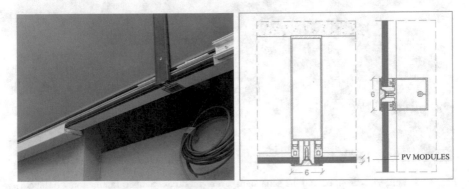

**Fig. 3.64** Left: detailed view of the modules mounting structure (Schüco International Italia Srl). Right: technical detail of Schüco ventilated façade system, re-drawn by Eurac research (Schüco International Italia Srl)

### DECISION-MAKING

Obtaining the LEED certification was one of the main reasons that lead the municipality to include a photovoltaic system, together with other project adjustments regarding the building construction features as well as the electric and hydraulic plants [3]. The photovoltaic plant was integrated on the building South façade, facing a large urban park and a busy street, highly visible to the community. It represents the municipality's commitment towards a sustainable environment (Arch. Gianluca Perottoni).

### PROCESS

In 2009, the municipality of Rovereto commissioned to Arch. Gianluca Perottoni (ViTre Studio) the Smart Lab project. The company Costruzioni Debiasi started the construction works. Later, Arch. Perottoni changed the project with the integration of the photovoltaic plant on the south-facing façade. He led the monitoring, accounting as well as the safety coordination during the construction phase. Schüco International Italia Srl was the main technical responsible for the design and the installation of the BIPV plant. In 2013, the building achieved the Silver level for the LEED sustainability rating system, according to the 'Leed Italia 2009 Nuove Costruzioni e Ristrutturazioni' protocol.

### LESSON LEARNT

The use of the microamorphous silicon technology creates a homogeneous aesthetically appealing surface, hiding the facade behind, which is made of glazed and opaque parts (Fig. 3.63, right). The microamorphous silicon modules are used together with an innovative mounting system that guarantees an effective thermal insulation effect and a high-quality architectural result (the joints outside visible are 60 mm wide) (Figs. 3.65, left and right). This consideration underlines how a continuous innovation of technologies can be essential to the increased diffusion of the BIPV systems. It should motivate the designers to research the best available products.

**Fig. 3.65** Left: view from behind the semi-transparent modules (Schüco International Italia Srl). Right: the thin-film modules create a high aesthetic quality building facade with a homogenized surface (Eurac research)

| BIPV SYSTEM DATA | |
| --- | --- |
| PV MODULES | Standard |
| SOLAR TECHNOLOGY | Microamorphous silicon |
| NOMINAL POWER | 10.8 kWp |
| SYSTEM SIZE | 130 m$^2$ |
| MODULE SIZE | 1,100 × 1,300 mm |
| ORIENTATION | South |
| TILT | 90° |

| BIPV SYSTEM COSTS | |
| --- | --- |
| TOTAL COST (€) | 85,000 |
| €/m$^2$ | 654 |
| €/kWp | 7,870 |

| PRODUCER DATA | |
| --- | --- |
| PRODUCER | Schüco International Italia Srl |
| ADDRESS | Via del Progresso 42, Padova (PD) |
| CONTACT | info@schueco.it |
| WEB | www.schueco.it |

## 3.2.16   Cable Car Station, Naturno (BZ)

| PROJECT DATA | |
|---|---|
| PROJECT TYPE | New construction |
| BUILDING FUNCTION | Industrial |
| INTEGRATION SYSTEM | Opaque cold façade, semi-transparent façade |
| LOCATION | Via Nazionale 2a, Laces (BZ) |

### BUILDING/SYSTEM INTEGRATION

#### Aesthetic integration

The glass envelopes that protect the valley (a) and the top (b) stations of the cable car in Naturno from the elements are made from semi-transparent BIPV modules (Fig. 3.66, and Figs. 3.67, left and right). The structures are located on a steep wooded slope of the Val Venosta in South Tyrol. The modules are integrated into both the lateral façades and on the southern pitch of the roofs. They were custom made in order to adjust the transparency and the size to the needs of the structures (Leitner Electro Srl).

**Fig. 3.66** Seilbahn Naturns BIPV system (Eurac research)

**Fig. 3.67** Left: view of the downstream unit semi-transparent glass roof (Eurac research). Right: view of the upstream unit semi-transparent glass façade (Leitner Electro Srl)

### Energy integration

With a nominal power of 19.4 kWp (a) and 30.4 kWp (b), the BIPV systems are estimated to produce 18,700 kWh/a and 24,800 kWh/a, respectively. The total annual energy production exceeds 50% of the demand (Leitner Electro Srl).

### Technology integration

The BIPV plants are made from polycrystalline glass–glass modules produced by Scheuten Optisol (P082136 K) (a) and EnergyGlass (EGP32ST/EGP48ST) (b). They are supported by a steel trusses system (Figs. 3.68, left and right). The BIPV modules are fixed to aluminium horizontal and vertical beams that hide the wiring system. They are naturally ventilated, due to the wide openings of the glazed structures.

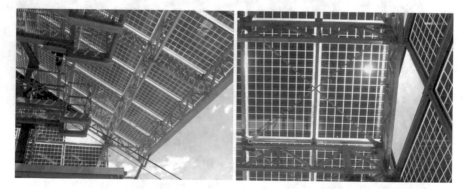

**Fig. 3.68** Left: steel trusses system supporting the BIPV plant (Leitner Electro Srl). Right: detailed view of the glass modules mounting structure (Leitner Electro Srl)

## DECISION-MAKING

The client Seilbahn Naturns GmbH wanted to build a roof for the ropeway station, in order to protect the technical system from the weather conditions. Architect Götsch envisioned to install a photovoltaic plant to lower the energy consumption and contribute to a sustainable environment. Full integration of the PV system into the roof makes a dual use (energetic and weather protection) of the panels possible. The aesthetic point of view was of great importance to the client. BIPV with semi-transparent modules was a convenient way to still get sufficient natural light into the building (Leitner Electro Srl) (Fig. 3.69, left).

**Fig. 3.69** Left: external view of the semi-transparent polycrystalline modules (Leitner Electro Srl). Right: the ropeway station roofs were optimally tilted (30°) for the photovoltaic integration (Eurac research)

## PROCESS

The client, Seilbahn Naturns GmbH, publicly opened a call for tenders. Architect Götsch was assigned to the building design. Leitner Electro Srl was responsible for the photovoltaic plant. The engineering department of Leitner Electro Srl and Pichler Stahlbau Srl planned and calculated the building structure and the PV plant together. The PV production was estimated with UNI 10-349 and a price calculation was made with a payback simulation. After the price calculation and the technical consulting, the client confirmed the installation of the plant. Stahlbau Pichler Srl built the construction and installed the glass modules, involving Vitralux Srl with the installation of the glass of the first plant. Leitner Electro Srl installed the other components of the plants. The first plant was completed in 2007, the second one in 2013 (Leitner Electro Srl).

## LESSON LEARNT

The flexibility of BIPV can increase its implementation into special places characterized by specific constraints (e.g. historical, environmental) as well as into different building typologies. This case study is an example of that since it is placed in a mountain context, on a steep wooded slope and it is a kind of building not commonly used for BIPV installations. The main building design did not need architectural alterations. The building roofs were optimally tilted (30°) for the application of a photovoltaic system (Fig. 3.69, right). Since it is an 'open configuration', no problems for internal comfort can occur, as it would have been in case of closed configuration.

| BIPV SYSTEM DATA | |
|---|---|
| PV MODULES | Custom made |
| SOLAR TECHNOLOGY | Polycrystalline silicon |
| NOMINAL POWER | 19.3 kWp (a)–30.4 kWp (b) |
| SYSTEM SIZE | 190 m$^2$ (a)–254 m$^2$ (b) |
| MODULE SIZE | Several |
| ORIENTATION | −40° (roofs), −130°/+50° (façades) |
| TILT | 30° (roofs), 90° (façades) |

| BIPV SYSTEM COSTS | |
|---|---|
| TOTAL COST (€) | 316,748 |
| €/m$^2$ | 713 |
| €/kWp | 6373 |

| PRODUCER DATA | |
|---|---|
| PRODUCER | Scheuten Solar Technology GmbH (a)– EnergyGlass Srl (b) |
| ADDRESS | Gelsenkirchen (D) (a)–Cantù (CO) (b) |
| CONTACT | contact@energyglass.eu (b) |
| WEB | www.scheutensolar.com (a)–www.energyglass.eu (b) |

## 3.2.17   Chamber of Commerce, Bolzano (BZ)

| PROJECT DATA | |
|---|---|
| PROJECT TYPE | New construction |
| BUILDING FUNCTION | Office |
| INTEGRATION SYSTEM | Opaque cold façade |
| LOCATION | Via Alto Adige 60, Bolzano (BZ) |

### BUILDING/SYSTEM INTEGRATION

#### Aesthetic integration

The Chamber of Commerce building is facing a central thoroughfare in the city of Bolzano. The PV system creates a dark vertical band integrated on a highly visible front, which is characterized by a variation of different façade materials (Fig. 3.70). Components like blasted stainless steel panels, parapets, structural glazing, coupled windows and daylighting elements (Frener & Reifer GmbH) form a single surface, creating a dynamic appearance marked by stringcourses (Fig. 3.71, left). The same combination of different materials and colours is reproduced inside the building (Fig. 3.71, right).

**Fig. 3.70** Chamber of Commerce BIPV system as a vertical dark band on the left side of the building (Arch. Wolfgang Simmerle)

**Fig. 3.71** Left: the modules are exposed on a heavily busy city path (Arch. Wolfgang Simmerle). Right: internal view: combination of different materials and colours, the same as the building external façade (Frener & Reifer GmbH)

**Energy integration**

The BIPV system installed in the façade is a small plant added to a bigger PV plant applied on the building's flat roof. Its electric output is fully self-consumed (Obrist GmbH). The produced energy is also used in the efficient thermal energy systems. The heating and cooling demand is supplied through a water-to-water heat pump, a free cooling system and natural gas-fueled condensing boilers (Energytech Srl). The building is CasaClima A+ certificated.

**Technology integration**

The 13 integrated photovoltaic modules Solarwatt (M234-108 GEG LK) are standard black glass–glass panels made of 108 monocrystalline cells. The 'glass-glass' technology, where solar cells are between two glass panes rather than of standard glass and plastic back-sheet setup, is considered as an extremely durable and resistant solution, with an optimal cells protection [4]. The PV plant is the top layer of an insulation element consisting of a metal sheet, an insulating layer and a concrete structure. An air gap (8 cm) is left between the panels and the metal sheet in order to allow natural ventilation. The modules are mounted as common curtain wall component. They are fixed to an aluminium frame made of cross and vertical beams, hiding the wiring system (Figs. 3.72, left and right).

**Fig. 3.72** Left: the photovoltaic energy output is visible on the façade surface (Eurac research). Right: detailed view of the monocrystalline modules (Eurac research)

## DECISION-MAKING

The integration of some photovoltaic modules on the highly visible building facade represents a symbol of the Province of Bolzano to the community, highlighting the local energy policy, which aims at sustainability through the exploitation of renewable energy (Frener & Reifer GmbH).

## PROCESS

In 2004, the Chamber of Commerce design process started. The Province of Bolzano commissioned to Arch. Wolfgang Simmerle with the general design of the building. Several specialists were involved. Energytech Srl and Industrie Team were responsible for the building's thermal and electric plants, respectively. Frener

& Reifer GmbH was chosen as technical responsible for the facade development, including BIPV system design and installation, together with Obrist GmbH team.

**LESSON LEARNT**

The PV modules are used as opaque facade panels. However, according to the designer (Frener & Reifer GmbH), the integration of semi-transparent modules in the large glazed surfaces of the building could have been an opportunity, assuming the function of a shading system (Fig. 3.73, right).

**Fig. 3.73** Left: the photovoltaic dark band splits the modern Chamber of Commerce from the close traditional building (Eurac research). Right: view of the south-facing glazed surfaces (Arch. Wolfgang Simmerle)

The close by building casts a partial shadow on the BIPV system (Fig. 3.73, left). Partial shadow of a PV system might cause severe power losses, since all cells and modules in an array are connected in series. The presence of bypass diodes in the PV modules can partially mitigate this problem, which anyway has to be carefully considered during the design phase.

| BIPV SYSTEM DATA | |
|---|---|
| PV MODULES | Standard |
| SOLAR TECHNOLOGY | Monocrystalline silicon |
| NOMINAL POWER | 3.3 kWp |
| SYSTEM SIZE | 30 m$^2$ |
| MODULE SIZE | 1,290 × 1,775 mm |
| ORIENTATION | South-west |
| TILT | 90° |

| BIPV SYSTEM COSTS | |
|---|---|
| TOTAL COST (€) | 26,800 |
| €/m$^2$ | 890 |
| €/kWp | 8,120 |

| PRODUCER DATA | |
| --- | --- |
| PRODUCER | Solarwatt GmbH |
| ADDRESS | Maria Reiche Straße 2A, Dresda (D) |
| CONTACT | info@solarwatt.net |
| WEB | www.solarwatt.net |

## 3.2.18   Castaneum Center, Velturno (BZ)

| PROJECT DATA | |
| --- | --- |
| PROJECT TYPE | New construction |
| BUILDING FUNCTION | Public |
| INTEGRATION SYSTEM | Opaque tilted roof |
| LOCATION | Piazza Silvius Magnago 1, Velturno (BZ) |

### BUILDING/SYSTEM INTEGRATION

#### Aesthetic integration

Castaneum centr is a mixed-use building in Velturno, a little village located on a sunlit hill of South Tyrol (Fig. 3.74). The building's facades are characterized by a modern appearance (Fig. 3.75, left). On the roof, a photovoltaic plant is integrated replacing the traditional roofing materials. It is inserted in the context of historical houses with their traditional gabled roofs (Fig. 3.75, right). The building preserves

**Fig. 3.74**  Castaneum Center BIPV roofing system (Arch. Albert Colz)

the tradition with a dark-pitched roof, with a decreased tilt angle, so that the BIPV modules, covering most of the available surface, are not visible from the street.

**Fig. 3.75** Left: modern building appearance (Arch. Albert Colz). Right: the innovative BIPV technology is used within a traditional urban context (Elektrostudio)

### Energy integration

The BIPV plant was sized to produce about 60,000 kWh, yearly. It was estimated to cover most of the building electricity demand as well as feeding a large amount of energy into the grid (Arch. Albert Colz). The solar power is not the only kind of renewable energy used in the building. The district heating plants of Velturno supplies the building's thermal demand. This wide use of renewable energy contributes to the CasaClima A certification of the building.

### Technology integration

193 PV modules (WINAICO WSP-M6 PERC Series) are integrated on the roof. This module typology is particularly suitable to be placed on high sunlit surfaces. It is characterized by the anti-PID (Potential Induced Degradation) technology, aimed to avoid a phenomenon that, enforced by high temperature and high level of humidity, can provoke permanent degradation of the p–n junctions [5]. The modules are installed on the concrete roof surface and follow the gable shape. They are fixed with aluminium crosswise clamps that create a gap between the PV panels and the concrete, allowing natural ventilation of the plant (Figs. 3.76, left and right).

**Fig. 3.76** Left: detailed view of the modules (Elektrostudio). Right: BIPV fixing system: the special crosswise clamps anchored to the concrete building roof are visible (Elektrostudio)

## DECISION-MAKING

The municipality of Velturno decided to apply photovoltaic technology in order to directly produce renewable energy on-site. Arch. Colz proposed to integrate the PV plant on the roof, so that it doesn't compromise the building aesthetic value and it is an energetically effective solution, not subject to shading between panels (Arch. Albert Colz).

## PROCESS

In 2008 the municipality of Velturno launched a call for a tender for the building design. Arch. Albert Colz won the competition, with a first project proposal that did not consider the photovoltaic implementation. Later, the PV was integrated into the plan. Elektrostudio was mainly responsible for the technical design and the installation of the PV system. Obrist GmbH was chosen as modules and system supplier. The building was completed in 2015.

## LESSON LEARNT

From the start of BIPV plant design phase, the aesthetic viewpoint played an important role. The photovoltaic surface was completed with optically matching black powder-coated panels, exactly aligned with the modules. Further, gang-ways—that partly correspond to the roof ridge—were installed for maintenance operations of the PV plant (Figs. 3.77, left and right).

**Fig. 3.77** Left: a space between the modules is left in order to permit through passage for the maintenance operations (Elektrostudio). Right: fake modules are installed, finishing the BIPV plant surface (Arch. Albert Colz)

Technically, an adjustment of the initial plan was needed. The PV panels were initially meant to be fixed to the building's steel structure directly, causing probable heat bridges through the concrete roof. The problem was solved with a design modification (Arch. Albert Colz). This highlighted the importance of performing detailed evaluations in designing complex components as a BIPV system.

| BIPV SYSTEM DATA | |
| --- | --- |
| PV MODULES | Standard |
| SOLAR TECHNOLOGY | Monocrystalline silicon |
| NOMINAL POWER | 50.2 kWp |
| SYSTEM SIZE | 316 m$^2$ |
| MODULE SIZE | 1,665 × 999 mm |
| ORIENTATION | South, West, East |
| TILT | 7°, 4° |

| BIPV SYSTEM COSTS | |
| --- | --- |
| TOTAL COST (€) | 130,000 |
| €/m$^2$ | 411 |
| €/kWp | 2,590 |

| PRODUCER DATA | |
| --- | --- |
| PRODUCER | Win Win Precision Technology Co., Ltd. |
| ADDRESS | Gongdao 5th Rd., Hsinchu City (Taiwan) |
| CONTACT | info@w-win.com.tw |
| WEB | www.wwpt.com.tw |

# References

1. CENELEC (2016) EN 50583—photovoltaics in buildings
2. Green Network Group (2013) Green planner magazine—Quartieri ecosostenibili: l'esperienza di Le Albere a Trento. [Online]. Available: https://www.greenplanner.it/2013/07/09/quartieri-ecosostenibili-lesperienza-di-le-albere-a-trento/
3. Municipality of Rovereto, "decree n° 2008 06.05/289-04 - Approvazione a tutti gli effetti della variante progettuale n° 2."
4. SolarwattGmbH (2016) The multi-generation modules—glass-glass modules
5. WINAICO Deutschland GmbH, "WINAICO data sheet"

# Chapter 4
# Conclusion

**Abstract** The call of case studies lead to the collection of more than 40 examples, a number which already shows how the vision of 'PV as building material' is slowly becoming a reality. Looking at the economic matter, it is shown that the BIPV systems capital cost lays in an acceptable range. BIPV resulted to be an absolutely viable solution from this point of view. Past and current incentive schemes has played a major role in boosting the use of PV in architecture, from an economic perspective as well as from the integration (technology, aesthetic and energy) point of view. BIPV is irrevocably set to play an essential role in the years to come, also thanks to the EU policy, and will have the opportunity to improve in all the aspects of its soul.

This book collects some of the best case studies of BIPV systems in the Trentino Alto Adige region.

The call of case studies lead to the collection of more than 40 examples, a number which already shows how the vision of 'PV as building material' is slowly becoming a reality.

This achievement is the result of a joint effort involving many players and institutions which have boosted the use of PV in architecture through the implementation of policies and incentive schemes on one side, and through the dissemination and information on the matter to increase the community awareness and technical knowledge on the other side.

At the national level, a major role was played by the feed-in tariff 'Conto Energia' which, from 2005 until 2013, has evolved to pursue more and more the 'marriage' between PV and architecture. In particular, the publication of the 'Catalogue of Photovoltaic Plants Integrated with Innovative Characteristics' in 2012 [1] has been an inspiring source for architects and designers to guide and boost the PV integration. At the local level, the policies of the two Provinces of Bolzano and Trento have often been at the forefront in boosting the exploitation of renewable energies and building energy efficiency through incentives, regulations and methodologies development.

The 18 presented case studies are analysed considering three aspects of the PV integration: **aesthetic** integration, **energy** integration and **technology** integration.

© Springer International Publishing AG, part of Springer Nature 2018
L. Maturi and J. Adami, *Building Integrated Photovoltaic (BIPV) in Trentino Alto Adige*, Green Energy and Technology,
https://doi.org/10.1007/978-3-319-74116-1_4

It is clear that, in order to succeed in the BIPV system design, all three aspects must be considered. For this reason, in our view, the 'I' of the acronym 'BIPV' should stand for 'integration' considering its triple meaning. This consideration leads to the conclusion that expertise in different branches is needed and different actors should be involved in the design and installation **process**. Regarding the described 18 case studies, more than 70 experts were interviewed to gather the necessary information, involving architect and engineering studio, façade manu-facturers, glass manufacturers, energy consultants, BIPV experts, PV installers, electricians, private and public building owners and real estate companies.

Figures 4.1 and 4.2 show a summary of the final-user **cost** of each BIPV system, looking at the same issue from two different perspectives (i.e. the 'PV' perspective and the 'BUILDING' perspective).

Figure 4.1 normalizes the BIPV systems cost to the nominal power (€/kWp), an indicator which is always used in the 'PV sector'. The cost of the analysed BIPV systems ranges from 2,500 to 8,300 €/kWp, with an average of around 5,500 €/kWp.

This variation can be ascribed to several factors, such as the type of technology integration, type of components and last but not least the construction year, since the PV cost has seen an impressive decrease in the last few years.

By comparing this data with standard not-integrated PV systems, we might conclude that the average 'BIPV' cost is around the double of a not-integrated one (considering a baseline cost of around 2,500 €/kWp, typical of small plants in the last recent years).

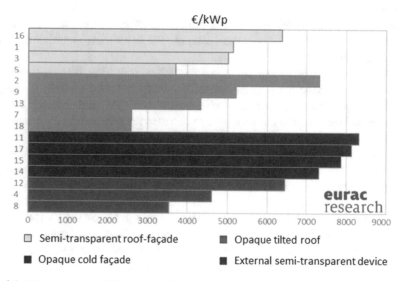

**Fig. 4.1** Final-user costs of the analysed BIPV systems, normalized to the system nominal power (cost information not available for some case studies)

In order to look at the economic matter from another perspective, the cost has been normalized to the envelope covered surface ($€/m^2$), thus using an indicator which is normally used in the 'BUILDING sector' (Fig. 4.2). The cost of the analysed BIPV systems ranges from 300 to 1,300 $€/m^2$, with an average of around 600 $€/m^2$. As mentioned above, this variation can be ascribed to several factors. In particular, this time a crucial role is played by the PV module efficiency. In fact, whether the PV cost/kWp is similar for different technologies (monocrystalline, polycrystalline and thin film), this is not the case for the PV cost/$m^2$ because more efficient modules (e.g. monocrystalline) have a higher peak power in a unit of area. On the other side, more efficient modules will produce more energy per unit area during their lifetime, thus paying back the initial cost. For this reason, looking at this indicator might be misleading but it is very useful to compare the BIPV system cost with standard building materials. It demonstrates that in fact the BIPV system capital cost lays in an acceptable range and it is even cheaper than some standard passive building materials (e.g. glazed curtain walls, stone and others). This, without even considering the payback time period, which ranges from 4 to 11 years for the presented case studies (this information was not available for all cases) and which is 'infinite' for standard passive solutions (without taking into consideration energy savings).

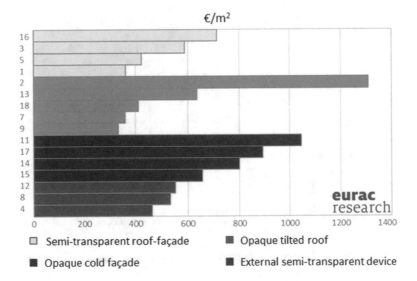

**Fig. 4.2** Final-user costs of the analysed BIPV systems, normalized to the envelope covered surface

As mentioned above, the cost variation is very much influenced by the construction year, since the PV cost has seen an impressive decrease in the last few years.

Figures 4.3 and 4.4 show the trend over the years of the final-user BIPV systems cost, considering the 'PV' and 'Building' perspective.

A clear decreasing trend is shown for the last decade (from 2004 to 2015) with values of $\sim 8,000$ €/kWp and $\sim 950$ €/m$^2$ in 2004 and of $\sim 3,300$ €/kWp and $\sim 400$ €/m$^2$ in 2015.

By comparing this data with standard not integrated PV systems, we might conclude that the 'BIPV' trend cost in 2015, corresponding to 3,300 €, is not too far from a ground-PV solution (considering a baseline cost of around 2,500 €/kWp, typical of small plants in the last recent years).

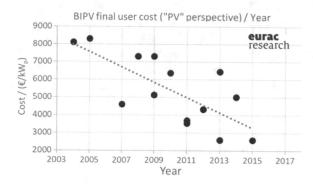

**Fig. 4.3** Final-user costs of the analysed BIPV systems per construction year, normalized to the system nominal power

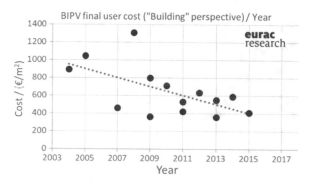

**Fig. 4.4** Final-user costs of the analysed BIPV systems per construction year, normalized to the envelope covered surface

This demonstrates that despite still the economic issue is perceived as a barrier for the widespread of BIPV systems [2], the use of PV in architecture is instead absolutely viable from this point of view.

Even if the 'Conto Energia' scheme has finished, it has paved the way to an irreversible process that cannot be stopped anymore.

The current support schemes at Italian level relies mainly on two measures: tax credit, which allows to recover 50% of the capital cost in 10 years, and the

'Scambio sul Posto' managed by GSE, which registers the energy produced by the PV and allows later consumption by the building.

The economic viability of BIPV systems is thus preserved, even if we can somehow read a conceptual shift in the way to reward it: with the 'Conto Energia' the aesthetic and technology integration was mainly boosted (through a higher contribution foreseen for 'innovative BIPV'), while, with the current schemes, the energy integration is mostly pursued, in order to maximize the energy match between the produced and consumed energy.

This energy integration will become more and more important to cope with the new ways buildings are conceived and their energy provision. In fact, also thanks to the EU policy oriented to promote the NZEB (nearly zero-energy buildings) concept and RES (renewable energy sources) exploitation [3, 4], buildings are becoming more than just stand-alone units using energy from the grid. They are becoming micro energy hubs consuming, producing, storing and supplying energy, thus transforming the EU energy market, shifting from centralized, fossil-fuel based, national systems towards a decentralized, renewable, interconnected and variable system.

In this context, PV integration is irrevocably set to play an essential role in the years to come and, learning from the experience gathered in realized projects, BIPV systems will have the opportunity to improve in all three aspects of its soul (technology, aesthetic and energy).

# References

1. GSE Gestore dei Servizi Energetici S.p.A. (2012) Catalogue of photovoltaic plants integrated with innovative characteristics
2. IEA Task 41 (012) Building integration of solar thermal and photovoltaics—barriers, needs and strategies, May-2012. [Online]. Available: http://task41.iea-shc.org/data/sites/1/publications/T41A1_Survey_FinalReport_May2012.pdf
3. SUPSI, "BIPV building integrated photovoltaic" [Online]. Available: http://www.bipv.ch/index.php/en/
4. IEA, "Database of innovative solar products for building integration" [Online]. Available: http://solarintegrationsolutions.org/

Printed in the United States
By Bookmasters